13th International Colloquium on Paratuberculosis (ICP 2016)

Nantes, France
20 - 24 June 2016

ISBN: 978-1-5108-4405-6

Printed from e-media with permission by:

Curran Associates, Inc.
57 Morehouse Lane
Red Hook, NY 12571

CURRAN ASSOCIATES INC.
proceedings
.com

Some format issues inherent in the e-media version may also appear in this print version.

Printed by Curran Associates, Inc. (2017)

International Association for Paratuberculosis
2015 Linden Drive
Madison, Wisconsin 53706-1102
USA

Phone: (608) 262-8457
Fax: (608) 265-6463

www.paratuberculosis.org

Additional copies of this publication are available from:

Curran Associates, Inc.
57 Morehouse Lane
Red Hook, NY 12571 USA
Phone: 845-758-0400
Fax: 845-758-2633
Email: curran@proceedings.com
Web: www.proceedings.com

TABLE OF CONTENTS

WELCOME MESSAGE

Dear colleagues,

We are pleased to welcome you to the 13th Colloquium of the International Association for Paratuberculosis held on June 20-23, 2016, in Nantes. Nantes is a dynamic and very pleasant city in the Western part of France, large enough to offer you a range of high standard facilities and activities, and small enough to make it easy to take the best out of it while you attend the conference or the days around it. We will make sure we share our best tips with you there.

As you all know, The International Colloquium on Paratuberculosis, periodically held under the auspices of the International Association for Paratuberculosis, has become a special meeting for the IAP members and for all leading researchers, livestock industry representatives, industry of animal health, veterinarians and public health authorities with an interest in the disease and its related issues.

Scientific sessions will focus on the latest advances in knowledge on the infection course and host response (including pathogenesis, immunology, and host genetics), on *Mycobacterium avium* subsp. *paratuberculosis* (including genomics, biology, and diversity of the pathogen), on diagnostic and detection (of the disease, the infection and the pathogen), on exposure and transmission (including epidemiology and modelling approaches to understand the disease dynamics in populations), on control strategies, and on public health and food safety aspects. Besides keynote speakers and oral presentations, time will be dedicated to specific sessions to present and discuss posters in a friendly and professional setting.

Our research group on Biology, epidemiology and risk analysis in animal health (BIOEPAR) fully supported by its institutes Oniris and INRA is honoured to receive the trust of the IAP to lead the organizational efforts to make the 13th colloquium an event of scientific interest and social enjoyment.
To do this, we have built an experienced and original group of French and international members to serve on the scientific committee.

For those of you who have a specific interest in getting in contact with stakeholders in France and sharing experience on the disease situation and control there, a stakeholders' day in French is organized on Friday June, 24, 2016.

The scientific content of ICP2016 is the result of your involvement. The organization of this conference has been made possible thanks to the support of our institutional partners and our sponsors. We also want to acknowledge the support of the Region Pays de la Loire and Nantes Metropole. Thanks to all, joint efforts will lead to a very successful conference.

We hope ICP 2016 will give you the best opportunities to share knowledge and network. We wish you enjoy fully the conference and take the best of your stay in Nantes.

On behalf of the Organizing and Scientific Committees

Chairs
Oniris – INRA
Christine Fourichon and Raphaël Guatteo

ORGANISING COMMITTEES

LOCAL ORGANISING COMMITTEE

Christine Fourichon (FR)	*christine.fourichon@oniris-nantes.fr*
Raphaël Guatteo (FR)	*raphael.guatteo@oniris-nantes.fr*
Gilles Foucras (FR)	*g.foucras@envt.fr*
Fabien Corbière (FR)	*f.corbiere@envt.fr*
Franck Biet (FR)	*franck.biet@tours.inra.fr*

IAP REPRESENTATIVES

Ramon Juste (Spain)	*rjuste@neiker.net*
Søren S. Nielsen (Denmark)	*saxmose@sund.ku.dk*
N. Arrigoni (Italy)	*norma.arrigoni@izsler.it*

IAP BOARD

http://www.paratuberculosis.net/

Officers of the association

President	Ramon Juste - Spain
Vice-President	Eiichi Momotani - Japan
Secretary	Raymond Sweeney - United States
Treasurer	Raymond Sweeney - United States
Editor-in-Chief	Søren S. Nielsen - Denmark

Board of directors

Richard Whittington - Australia
Jeroen DeBuck - Canada
Gregers Jungersen - Denmark
Christine Fourichon - France
Heike Koehler - Germany
Shoorvir Singh - India
Peter Mullowney - Ireland
Norma Arrigoni - Italy
Victor Rutten - Netherlands
Frank Griffin - New Zealand
Joseba Garrido - Spain
Karen Stevenson - United Kingdom

SCIENTIFIC COMMITTEE

INFECTION COURSE AND HOST RESPONSE (INCLUDING HOST GENETICS)

Gilles Foucras (France)	INP-ENVT	*g.foucras@envt.fr*
Jeroen Debuck (Canada)	Calgary University	*jdebuck@ucalgary.ca*
Richard Whittington (Australia)	Sydney University	*richard.whittington@sydney.edu.au*
Gregers Jurgensen (Denmark)	Copenhagen University	*grju@vet.dtu.dk*

PATHOGEN GENOMICS, DIVERSITY, AND BIOLOGY

Franck Biet (France)	INRA UMR ISP	*franck.biet@tours.inra.fr*
John Bannantine (USA)	USDA-ARS	*john.bannantine@ars.usda.gov*
Karen Stevenson (UK)	Moredun Institute	*karen.stevenson@moredun.ac.uk*
Marcel Behr (Canada)	Calgary University	*marcel.behr@mcgill.ca*

DIAGNOSTIC & DETECTION

Allen Roussel (USA)	Texas A&M University	*ARoussel@cvm.tamu.edu*
Heike Kohler (Germany)	FLI-Munich	*heike.koehler@fli.bund.de*
Georges Caldow (UK)	SRUC	*George.Caldow@sac.co.uk*
Matteo Ricchi (Italy)	Izsler Institute	*matteo.ricchi@izsler.it*

EXPOSURE AND TRANSMISSION

Fabien Corbière (France)	INP-ENVT	*f.corbiere@envt.fr*
Cord Heuer (New-Zealand)	Massey University	*C.Heuer@massey.ac.nz*
Maarten Weber (The Netherlands)	GD-the netherlands	*m.weber@gdanimalhealth.com*
Karsten Donat (Germany)	Thüringer Tierseuchenkasse	*kdonat@ thueringertiseuchenkasse.de*

SPECIAL WORKSHOP : FOCUS ON MODELLING TO UNRAVEL THE DYNAMICS OF INFECTION AND SPREAD

Pauline Ezanno (FR)	INRA	*pauline.ezanno@oniris-nantes.fr*

ONE HEALTH AND CONTROL PLANS

Raphaël Guatteo (France)	ONIRIS	*raphael.guatteo@oniris-nantes.fr*
Joseba Garrido (Spain)	Neiker Institute	*jgarrido@neiker.net*
Karin Orsel (Canada)	Calgary University	*karin.orsel@ucalgary.ca*

13 ICP IAP AWARDS

Richard S. Merkal Awards

Hannah B Pooley	Australia
Caroline Ritter	Canada

Helping Hand Awards

Kundan Kumar Chaubey	India
Sujata Hariharan	India
Kamal Raj Acharya	Nepal
Ana Carolina Silva Faria	Brazil
Marat Kuibagarov	Kazakhstan

COLLOQUIUM GENERAL INFORMATION

CONFERENCE VENUE

La Cité Nantes Event Center
5, rue de Valmy
44041 Nantes, France
All the scientific sessions, the IAP General meeting, the IAP Board of directors and the ParaTB Forum will be held at the conference venue.

POSTER PRESENTATIONS

In preparing your poster presentation please use the following guidelines :

- Size: max A0 : 841 x 1189 mm / 33" x 47"– in portrait orientation
- Written in English
- If figures are included, these should be clearly labelled and legible

Please check in with the onsite Secretariat Registration Desk at the La Cité Nantes Events Centre when you pick up your registration materials. They will direct you to the poster area for your session.

Posters will be displayed in groups by Session. Your poster session and number can be found by checking the Author index in the Program Book. Authors may begin putting up their posters on Monday, 20 June 2016 after the registration. You can pick up the supplies you need to fix your poster to the boards from the Registration Desk. Also, check with them if you need further assistance regarding your poster.

Posters will be available for viewing during all registration opening times. We ask that all posters remain on display until after the Synopsis on Thursday June 23rd, at which time authors may then collect their posters. Specific time is dedicated to a poster session every day. We strongly suggest that authors be at their posters to discuss their work and answer questions during the poster session on the day of your session.

Printing and transport of the poster is at author's charge and care.

ORAL PRESENTATION

In preparing your oral presentation please remember to :

- Include clear objectives
- Have a legitimate study design to investigate the stated objectives
- Discuss valid, substantiated conclusions
- Use clear and comprehensible language
- Create relevant graphs, photos, tables using standard software (Microsoft Office Power Point 2000/2003/2007/2010)
- Save slide data (.PPT or PPTX) on a USB flash memory drive (previously checked using your own pc)

NOTE: There is no system for editing the slide in our session room.

Receiving the slide data:
Please bring your USB flash drive to the slide desk in the session room at least 3 hours before the beginning of the relevant session. A technician will upload your slide presentation and return your flash drive. The copied slides will be erased after each speech.

Please note it is not possible to use your own pc. Authors must present themselves to the Chairperson of the session at least 15 minutes before the session starts in order to acquaint themselves with the equipment. Length of the oral presentations will be a total of 15 minutes (10 min. presentation + 5 min. discussion). Authors are kindly requested to respect the indicate timings.

You can check the timing of the oral sessions, and your presentation, on the website (under Scientific Program) and also in the printed Program book found in your Colloquium bags.

CERTIFICATE OF ATTENDANCE

A Certificate of Attendance will be provided to each delegate at the conference.

DISCLAIMER

The Colloquium Committee reserves the right to change the scientific program at any time without notice. Please note this program is correct at time of printing.

NAME BADGES

For security purposes, and to facilitate collaboration, delegates, speakers, sponsors, and exhibitors are asked to wear their name badges to all sessions.
If you misplace your name badge, please enquire at the Registration Desk to organize a replacement.

TICKETS

No entrance allowed at the Colloquium Dinner Thursday, June 23rd, without entrance ticket, if purchased.

PROGRAM

Every effort has been made to produce an accurate Program. If you are presenting at the Colloquium please ensure you double check your presentation time as indicated in this Program.

REGISTRATION DESK

The Registration Desk is located at the main entrance. It will be open at the following times :

Sunday, 19 June 2016	14.00 -19.00
Monday, 20 June 2016	8.00 -19.00
Tuesday, 21 June 2016	8.00 -17.00
Wednesday, 22 June 2016	8.00 -18.00
Thursday, 23 June 2016	8.00 -17.00
Friday, 24 June 2016	9.30 -17.00

SPEAKERS

Please ensure you are available in the conference room at least 15 minutes before the start of the session. Speakers, please visit the slide desk located at the back of the conference room at least 3 hours prior to the start of your session.

INTERNET

There is free wifi in the conference center. Look in your conference bags for login information or check with the Registration desk.

MOBILE PHONES

Please switch off your mobile phones during sessions. Thank you!

PARKING

At the La Cité Nantes Events Centre.

PROCEEDINGS

The proceedings are on the USB storage device you received in your registration materials as well as currently available on the Colloquium website (www.colloque.inra.fr/icp2016).
They will also be available on the IAP website (www.paratuberculosis.info) after the Colloquium under the Publications tab.

SPECIAL DIETARY REQUIREMENTS

If you have advised the Colloquium Secretariat of special dietary requirements, please speak to a member of the catering staff at catered events to receive your special meal.

SMOKING

Please be advised that smoking is not permitted in any public buildings.

BANCOMAT

Automatic Teller Machines (ATM)/Bancomat at the post-office or banks, 200 m from the congress center.

TRANSPORTATION IN NANTES

The conference venue is located in the city centre and is easily accessible from many hotels within walking distance. There is a wide range of choice by public transportation (tram, bus). A weekly pass valid for all public transportation in Nantes can be obtained at the vending machines on tram and busway stations (one is facing the conference centre), and at the railway station. Shared bicycles can also be rented and stored at many points in the city centre (look for the map of "Bicloo" stations, one is facing the conference centre).

CURRENCY AND MONEY EXCHANGE

The monetary unit in France is the Euro. Currency can be exchanged at the airport, at banks or from cash dispensers. All major credit cards such as MasterCard, Visa and American Express are accepted at most hotels, restaurants, shops and at the conference centre.

ELECTRICITY SUPPLY

France's electrical system is 220 V, 50 Hz and the outlets are of the two-round-pin type.

CLIMATE

Nantes has a mild oceanic climate. June is a pretty warm month with lows around 12°C and highs around 22-24°C. It is also relatively dry (average 40 mm precipitation in the month) and sunny (average 210 h). Under the influence of the ocean, there may be showers when a low is passing through. In Nantes we then generally say that we see the sun several times per day.

HEALTH AND INSURANCE

Participants are requested to arrange their own insurance for health, travel and property. The Organizing Committee will not accept any liability for personal injuries, loss or damage of property.

SOCIAL ACTIVITIES

GET TOGETHER DRINK

Sunday, June 19th

The Brady's pub – 22, allée Commandant Charcot, Nantes

Starting at h. 19.30

Free for registered delegates and registered accompanying persons.

At a 5-minute walk from the conference venue, get together, meet old friends and make new ones before the conference starts. Light drinks will be served. After a good chat you will still have plenty of time to find a restaurant in the surroundings for dinner.

COLLOQUIUM DINNER

Thursday, June 23rd

Château des Ducs de Bretagne

4, place Marc Elder, Nantes

From h 19.00 to 23.30

No participation allowed without a ticket.

After a busy and stimulating week, grasp the taste of being a privileged guest of a famous castle of the Loire Valley. Discover and learn more about French and Brittany history before you share a typical gourmet dinner and a lively and relaxing evening with your Colloquium friends.

WINE TASTING

Monday, June 20th

At the Conference venue,
from h. 17.30 to 19.30

Free for registered delegates and registered accompanying persons.

To prolong discussions started during the first poster session, that day will end with a fabulous wine tasting evening. You will challenge and compare know-how from major French wine-producing areas. A wine expert has selected a choice of wines (white and red) with varied flavour profiles. Because France is also "the country of 365 cheeses", a selection of cheeses that perfectly match with each wine will be served to make your exploration an even greater culinary pleasure. Our expert will give you all the information you need on tasting and on the products.

Tips for your free time: don't miss!

On Tuesday, June 21st, a nationwide cultural event "La Fête de la Musique" (The Music Day). A broad range of musical events, more or less amateur, of many different types and styles will be offered in many places all across the city of Nantes.

The exhibition «Machines de l'Ile» with the Giant mechanical walking Elephant, and the Marine Worlds Carousel, granted "Most original attraction of the World" in 2014. Take a unique chance to get a ride on the elephant while it walks along the Loire River, or just walk alongside.
http://www.lesmachines-nantes.fr/en/

COLLOQUIUM PARTNERS & SPONSORS

INSTITUTIONAL PARTNERS

Oniris — École Nationale Vétérinaire, Agroalimentaire et de l'Alimentation Nantes Atlantique

INRA — SCIENCE & IMPACT

International Association for Paratuberculosis

GDS France

Cniel

anses — agence nationale de sécurité sanitaire alimentation, environnement, travail

INSTITUT DE L'ELEVAGE — idele.fr

Région PAYS DE LA LOIRE

Nantes Métropole — COMMUNAUTÉ URBAINE

PLATINUM SPONSORS

CZV — CZ VETERINARIA

IDEXX

SILVER SPONSORS

IDvet — Innovative Diagnostics

biosellal

ThermoFisher SCIENTIFIC

ADIAGENE

BioX Diagnostics

BRONZE SPONSORS

MERIAL — UNE SOCIETE SANOFI

PROGRAM AT A GLANCE

SUNDAY 19th	
14.00	Registration opens
14.00-18.00	ParaTB FORUM (by invitation)
15.00-19.00	IAP Board of Directors
MONDAY 20th	
8.30-9.00	Conference opening
SESSION 1 - INFECTION COURSE	
9.00-9.45	**Key note Lecture : Rasmus Mortensen**
PATHOGENESIS & EXPERIMENTAL INFECTION	
9.45-10.15	Oral Sessions
10.15-10.45	*Coffee break*
10.45-12.15	Oral Sessions
12.15-13.15	*Lunch - 1 hour*
MAP PHENOTYPES	
13.15-14.45	Oral Sessions
14.45-15.15	*Tea break*
GENOMIC AND GENETICS OF THE HOST	
15.15-16.15	Oral Sessions
16.15-17.30	Poster Sessions
17.30-19.00	Poster Session - Wine Tasting

TUESDAY 21st	
SESSION 2 - MAP GENOMICS & DIVERSITY	
8.30-9.15	**Key note Lecture : Pr S. Gordon**
MAP DIVERSITY & BIOLOGY	
9.15-10.15	Oral Sessions
10.15-10.45	*Coffee break*
MAP GENOMICS AND MOLECULAR EPIDEMIOLOGY	
10.45-12.15	Oral Sessions
12.15-13.15	*Lunch - 1 hour*
13.15-14.45	Poster Sessions
14.45-15.15	*Tea break*
SESSION 3 - DIAGNOSTIC AND DETECTION	
IMMUNITY & DIAGNOSTIC	
15.15-16.00	**Key note Lecture : Pr S. McKenna**
16.00-17.00	Oral Sessions

WEDNESDAY 22nd	
SESSION 3 - DIAGNOSTIC AND DETECTION	
INFORMATIVE VALUE OF DIAGNOSTIC TESTS	
8.30-9.30	Oral Sessions
ENVIRONMENTAL DIAGNOSTIC	
9.30-10.15	Oral Sessions
10.15-10.45	*Coffee break*
10.45-12.15	Poster Sessions in parallel with workshop modelling
12.15-13.15	*Lunch - 1 hour*
13.15-13.45	Oral Sessions
NEW APPROACHES	
13.45-14.45	Oral Sessions
14.45-15.15	*Tea break*
SESSION 4 - EXPOSURE AND TRANSMISSION	
15.15-16.00	**Key note Lecture : Pr C. Heuer**
16.00-17.00	Oral Sessions
17.00-18.00	IAP General meeting

THURSDAY 23rd	
SESSION 5 - ONE HEALTH & CONTROL PLANS	
8.30-9.15	**Key note Lecture : L. Waddell**
MODELLING	
9.15-9.45	Oral Sessions
VACCINATION	
9.45-10.45	Oral Sessions
10.45-11.15	*Coffee break*
CONTROL PLANS	
11.15-12.45	Oral Sessions
12.45-13.45	*Lunch - 1 hour*
13.45-14.45	Poster Sessions
14.45-15.15	*Tea break*

PROGRAM AT A GLANCE

THURSDAY 23rd	
SESSION 5 - ONE HEALTH & CONTROL PLANS	
FARMERS PERCEPTION & ATTITUDES	
15.15-16.15	Oral Sessions
16.15-17.00	**Synopsis on the conference and closure**
19.00-23.30	*Conference dinner in the Duchesse Anne Castle*

FRIDAY 24th	
STAKEHOLDERS' DAY - JOURNEE PROFESSIONNELLE	
10.00-10.10	**Introduction journée** - C. Fourichon, Professionnel Réseau GDS
10.15-11.30	Mise en place de la vaccination : de l'autorisation à la mise en œuvre
11.30-12.00	*Pause café* - Posters
12.00-13.15	Les nouveautés en matière de paratuberculose
13.15-14.30	*Repas / Café* - Posters
14.30-16.00	Niveaux de risque acceptables lors de mouvements / introduction d'animaux
16.00-17.00	Ce qui démarre / va se faire en termes de plans de maîtrise
	Clôture : Mise en œuvre de la coordination nationale (GDS France) et les actions prioritaires prévues (ACERSA, Professionnel GDS France)

FINAL PROGRAM

MONDAY 20 JUNE 2014

from 9.00 to 13.15

INFECTION COURSE

9.00 **Key note Lecture: Rasmus Mortensen**

PATHOGENESIS & EXPERIMENTAL INFECTION

9.45 O-01.1 **IN VITRO EXPERIMENTAL WORK AND AGENT-BASED MODELING OF GRANULOMA FORMATION CAUSED BY MYCOBACTERIUM AVIUM SUBSP. PARATUBERCULOSIS**
Shigetoshi Eda[1], J. Hunter Rice[1], Alyson Marek[1], Sarah M. Alouani[1], Judy Day[2]
[1]Department of Forestry, Wildlife and Fisheries, University of Tennessee Institute of Agriculture. [2]Department of Mathematics, University of Tennessee, Knoxville, USA

10.00 O-01.2 **EARLY INTERACTIONS BETWEEN BOVINE MONOCYTE DERIVED MACROPHAGES AND MYCOBACTERIUM AVIUM SUBSP. PARATUBERCULOSIS**
Heather Mathie[1], Joanne Stevens[1], Elizabeth Glass[1], Jayne Hope[1]
[1]The Roslin Institute, The University of Edinburgh, Edinburgh, Scotland

10.15 Coffee break

10.45 O-01.3 **RABBIT GUT MICROBIOTA MODIFICATIONS INDUCED BY MYCOBACTERIUM AVIUM SUBSPECIES PARATUBERCULOSIS INFECTION AND DIET CHANGES**
Rakel Arrazuria[1], Elena Molina[1], Iker A.Sevilla[1], Miriam Serrano[1], Hooman Derakhshani[2], Ehsan Khafipour[2,3], Valentín Perez[4], Joseba Garrido[1], Ramon A. Juste[1,5], Natalia Elguezabal[1]
[1]NEIKER-Tecnalia, Basque Institute for Agriculture Research and Development, Derio, Bizkaia, Spain; [2]Department of Animal Science, University of Manitoba, Winnipeg, MB, Canada; [3]Department of Medical Microbiology, University of Manitoba, Winnipeg, MB, Canada; [4]Department of Animal Health, Faculty of Veterinary Medicine, University of León, Spain; [5]SERIDA, Department of Agriculture of the Regional Government of the Principality of Asturias, Deva, Asturias, Spain

11.00 O-01.4 **QUANTIFYING RESILIENCE TO ARTIFICIAL SYSTEMIC MAP INFECTION USING HISTOPATHOLOGY AND THE IDENTIFICATION OF HEPATIC EPITHELIOIDE MACROPHAGE (HEM) MICRO-GRANULO MAS FROM SERIAL LIVER BIOPSIES IN SHEEP.**
Stefan Lindsay Smith[1], Professor Peter Wilson[1], Professor Dave West[1], Professor Cord Heuer[2], Associate Professor Paul Chambers[1]
[1]Institute of Veterinary, Animal and Biomedical Sciences, Massey University, Palmerston North, New Zealand; [2]The Epi Centre, Massey University, palmerston North, New Zealand

11.15 O-01.5 **EXPERIMENTAL INFECTION OF CATTLE WITH MYCOBACTERIUM AVIUM SUBSP. PARATUBERCULOSIS: RESULTS FROM THE FIRST TWO YEARS**
Louise Britton[1], Colm Brady[2], Joseph Cassidy[1], Jim O'Donovan[3], Nola Leonard[1], Eamon Gormley[1], Ronan Shaughnessy[1], Stephen V. Gordon[1], Bryan Markey[1]
[1] School of Veterinary Medicine, University College Dublin, Ireland.[2] Department of Agriculture, Food and the Marine, Central Veterinary Research Laboratory, Backweston, Kildare, Ireland.[3] Department of Agriculture, Food and the Marine, Regional Veterinary Laboratory, Model Farm Road, Cork, Ireland

11.30 O-01.6 **IDENTIFICATION AND EVALUATION OF PROGNOSTIC BIOMARKERS FOR MYCOBACTERIUM AVIUM SUBSP. PARATUBERCULOSIS INFECTION**
Hong-Tae Park[1], Hyun-Eui Park[1], **Han Sang Yoo**[1,2]
[1]Department of Infectious Diseases, College of Veterinary Medicine, Seoul National University, Seoul, 08826, Republic of Korea. [2]Institute of Green-Bio Science and Technology, Seoul National University, Pyeongchang, 25354, Republic of Korea

11.45 O-01.7 **Differential systemic and local cytokine-responses after experimental infection of goats with Mycobacterium avium subsp. paratuberculosis (MAP) or Mycobacterium avium subsp. hominissuis (MAH)**

Heike Köhler[1], Anja Zigan[1], Jan Schinköthe[1], Petra Möbius[1], Elisabeth M. Liebler-Tenorio[1]
[1]Institute of Molecular Pathogenesis, Friedrich-Loeffler-Institut, Jena, Germany

12.00 O-01.8 **The importance of cholesterol in MAP infection of ruminants**

Matt Johansen[1], de Silva K[1], Plain KM[1], Begg D[1], Biet F[2], Whittington RJ[1], Purdie AC[1].
[1]The University of Sydney, Camden, Australia. [2]Institut National de la Recherche Agronomique, Nouzilly, France

12.15-13.15 Lunch

from 13.15 to 19.00

13.15 O-01.9 **Case definitions for paratuberculosis – the need for standards**

Richard Whittington[1], Kumudika de Silva[1], Douglas Begg[1], Auriol Purdie[1], Navneet Dhand[1], **Karren Plain**[1]
[1]School of Life and Environmental Sciences, Faculty of Veterinary Science, The University of Sydney, Australia

13.30 O-01.10 **Effects of vaccination before or after M. a. subsp paratuberculosis (Map) experimental infection in goats**

Royo M[1], Fuertes M[1], Fernández M[1], Sevilla IA[2], Arrazuria R[2], Castaño P[1], Ferreras MC[1], Benavides J[1], Elguezabal N[2], Pérez V[1]
[1] Dpt Sanidad Animal, Instituto de Ganadería de Montaña (CSIC-ULE), Facultad de Veterinaria, Universidad de León, León, Spain.[2] Dpt of Animal Health, NEIKER-Instituto Vasco de Investigación y Desarrollo Agrario. Derio, Bizkaia, Spain

13.45 O-01.11 **Phenotypic diversity in the immune response against Mycobacterium avium subsp. paratuberculosis in MAP-infected dairy cows**

Meredith Frie[1], Kelly Sporer[2], Paul Coussens[2]
[1]Cell and Molecular Biology Program, Michigan State University, East Lansing, MI, USA.[2]Department of Animal Science, Michigan State University, East Lansing, MI, USA

14.00 O-01.12 **Paratuberculosis vaccinated cattle lifespan and unspecific protection**

Juste RA[1,2], Vazquez P[1,3], Geijo MV[1], Serrano M[1], Elguezabal N[1], Molina E[1], Sevilla I[1], Alonso Hearn M[1], Perez V[4], Garrido JM[1]
[1] NEIKER-Tecnalia, Basque Institute for Agricultural Research and Development, Derio, Bizkaia, Spain.
[2] SERIDA, Regional Service of Agricultural Research and Development, Villaviciosa, Asturias, Spain. [3] SALUVET, Department of Animal Health, Faculty of Veterinary Sciences. Complutense University of Madrid (CUM).
[4] Department of Animal Health. Veterinary School. University of Leon. Leon, Spain

14.15 O-01.13 **Early Markers for Infection and Vaccination Against Johne's Disease**

Adel M. Talaat[1], Aubrey Berry[1]
[1]University of Wisconsin-Madison, WI, USA

14.30 O-01.14 **Evidence and detection of shedding patterns of Mycobacterium avium susbspecies paratuberculosis in dairy cattle**

R. Guatteo[1], P. Blanquefort[2], A.Delafosse[2], A. Joly[2], E. Meens[2], C. Fourichon[1]
[1]Oniris, UMR Oniris-INRA 1300 BioEpAR, Nantes, France. [2]GDS Grand Ouest, France

14.45-15.15 Coffee Break

GENOMIC AND GENETICS OF THE HOST

15.15 O-01.15 **COMPOSITE IMMUNE BIOMARKERS ASSOCIATED WITH RESISTANCE OR SUSCEPTIBILITY TO JOHNE'S DISEASE: DATA FROM A DEER MODEL**

Frank T Griffin[1], Colin G Mackintosh[2], Rory O'Brien[1], Simon Liggett[1], Liam Brennan[1]
[1]Disease Research Laboratory, University of Otago, Dunedin, New Zealand.[2]Invermay Agricultural Centre, Mosgiel, New Zealand

15.30 O-01.16 **WHOLE GENOME ASSOCIATION ANALYSIS OF RESISTANCE / SUSCEPTIBILITY TO PARATUBERCULOSIS IN FRENCH HOLSTEIN AND NORMANDE CATTLE**

Sanchez MP[1], Guatteo R[2], Davergne A[3], Grohs C[1], Capitan A[1,4], Blanquefort P[5], Delafosse A[6], Joly A[7], Ngwa-Mbot D[8], Biet F[9], Fourichon C[2], Boichard D[1]
[1]GABI, INRA, AgroParisTech, Jouy en Josas, France. [2]UMR Oniris-INRA 1300 BIoEpAR, Nantes, France. [3]Groupement de Défense Sanitaire Haute-Normandie, Rouen, France. [4]Allice, Paris, France. [5]Groupement de Défense Sanitaire Pays de la Loire, Trelazé, France. [6]Groupement de Défense Sanitaire Orne, Alençon, France. [7]Groupement de Défense Sanitaire Bretagne, Vannes, France. [8]GDS France, Paris, France. [9]ISP, INRA, Nouzilly, France

15.45 O-01.17 **GENE EXPRESSION ANALYSIS IDENTIFIES GENES DIFFERENTIALLY REGULATED IN BOVINE AND OVINE MODELS OF MYCOBACTERIUM AVIUM SUBSPECIES PARATUBERCULOSIS EXPOSURE**

Auriol Purdie[1], Kumudika de Silva[1], Douglas Begg[1], Karren Plain[1], Richard Whittington[1]
[1]School of Life and Environmental Sciences, Faculty of Veterinary Science, The University of Sydney, Australia

16.00 O-01.18 **OVINE MACROPHAGE-MEDIATED KILLING OF MYCOBACTERIUM AVIUM SUBSPECIES PARATUBERCULOSIS IN VITRO IS ENHANCED BY VACCINATION**

Hannah B Pooley[1]**,** Kumudika de Silva[1], Auriol C. Purdie[1], Douglas J. Begg[1], Richard J. Whittington[1], Karren M. Plain[1]
[1]School of Life and Environmental Sciences, Faculty of Veterinary Science, The University of Sydney, Camden, NSW, Australia

16.15 POSTER SESSIONS

17.30-19.00 POSTER SESSIONS + WINE TASTING

TUESDAY 21 JUNE 2014

from 8.30 to 13.15

MAP GENOMICS & DIVERSITY

8.30 **Key note Lecture: Pr S. Gordon**

MAP DIVERSITY & BIOLOGY

9.15 O-02.1 **MYCOBACTERIUM AVIUM SUBSP. PARATUBERCULOSIS FATTY-ACID COMPOSITION MIGHT DETERMINE ITS INTERACTION WITH BOVINE MACROPHAGES AND INFLUENCE THE INTRACELLULAR SURVIVAL OF THE BACILLUS.**

Marta Alonso-Hearn[1], Naiara Abendaño[1], Amparo Rubirar[2], Rosa Aznar[2], Ramon A. Juste[1,3]
[1]Neiker-Tecnalia, Basque Institute for Agricultural Research and Development, Derio, Bizkaia, Spain. [2]Spanish Type Culture Collection (CECT), Identification and Characterization Service of Microbial Isolates, Paterna, Valencia, Spain. [3]SERIDA, Department of Agriculture of the Regional Government of the Principality of Asturias, Deva, Asturias, Spain

9.30 O-02.2 **Genes essential for Mycobacterium avium subsp. paratuberculosis survival and persistence following oral infection of dairy calves**

Jeroen De Buck[1], Amanda Mirto[1], Joyce Wang[2], Marcel Behr[2], Herman W. Barkema[1]

[1]Department of Production Animal Health, University of Calgary, Faculty of Veterinary Medicine, Department of Production Animal Health. 3330 Hospital Drive NW, T2N 4N1, Calgary, Alberta, Canada. [2]McGill University, Faculty of Medicine, Department of Microbiology, McGill University Health Centre, 1001 boul Décarie, Montréal, Quebec H4A 3J1, Canada

9.45 O-02.3 **Cell Wall lipopeptides of Mycobacterium avium: New insights from genomics analysis**

John P. Bannantine[1], Gilles Etienne[2,3], Françoise Laval[2,3], Anne Lemassu[2,3], Mamadou Daffé[2,3], Darrell O. Bayles[1], Christelle Ganneau[4], Maxime Branger[5], Thierry Cochard[5], Sylvie Bay[4] and **Franck Biet**[5]

[1]USDA-Agricultural Research Service, National Animal Disease Center, Ames, Iowa, U.S.A. [2]Institut de Pharmacologie et de Biologie Structurale, Université de Toulouse, CNRS, UPS, FranceCNRS, Institut de Pharmacologie et de Biologie Structurale, 31000 Toulouse, France. [3]Université de Toulouse, UPS, IPBS, 31000 Toulouse, France. [4]Institut Pasteur, Unité Chimie des Biomolécules, CNRS UMR 3523, 75724 Paris Cedex 15, France. [5]INRA, UMR1282, Infectiologie et Santé Publique, F-37380 Nouzilly, France

10.00 O-02.4 **Proteins of Mycobacterium avium subspecies paratuberculosis that Bind and Hydrolyze Peptidoglycan**

John Bannantine[1], Judith R. Stabel[1], Cari K. Lingle[2], Philip R. Adam[3], Kasra X. Ramyar[4], William J. McWhorter[4], William D. Picking[3], Brian V. Geisbrecht[4]

[1]USDA-Agricultural Research Service, National Animal Disease Center, Ames, IA. [2]School of Biological Sciences; University of Missouri-Kansas City, Kansas City, MO. [3]Department of Microbiology & Molecular Genetics; Oklahoma State University, Stillwater, OK. [4]Department of Biochemistry and Molecular Biophysics, Kansas State University, Manhattan, KS, USA

10.15-10.45 Coffee Break

MAP GENOMICS AND MOLECULAR EPIDEMIOLOGY

10.45 O-02.5 **Construction and characterization of a lprG mutant of Mycobacterium avium subsp paratuberculosis K10**

Viale Mariana Noelia[1], Park Kun Taek[2], Imperiale Belén[1], Alonso Natalia[1], Colombatti Olivieri María Alejandra[1], Davis William[2], Romano María Isabel[1], **Santangelo María de la Paz**[1]

[1]Instituto de Biotecnologia, Instituto Nacional de Tecnología Agropecuaria, Hurlingham, Buenos Aires 1686, Argentina. [2]Department of Veterinary Microbiology and Pathology, College of Veterinary Medicine, Washington State University, Pullman, WA 99164, USA

11.00 O-02.6 **New WGS data of MAP-S strain JIII-386 from Germany and comprehensive genome analysis unveiled novel CDS, regulatory elements and polymorphism regions**

Möbius P.[1], Hölzer M.[2], Felder M.[3], Nordsiek G.[4], Groth M.[3], Köhler H.[1], Reichwald K.[3], Platzer M.[3], Marz M.[2]

[1]Institute of Molecular Pathogenesis, Friedrich-Loeffler-Institut (Federal Research Institute for Animal Health), Jena, Germany. [2]Faculty of Mathematics and Computer Science, Friedrich-Schiller-University Jena, Germany. [3]Leibniz Institute on Aging - Fritz-Lipmann-Institute, Jena, Germany. [4]Department of Genome Analysis, Helmholtz Centre for Infection Research, Braunschweig, Germany

11.15 O-02.7 **Phylogeny of New Zealand strains of Mycobacterium avium paratuberculosis by whole genome sequencing**

Marian Price-Carter[1], Des Collins[2], Geoff de Lisle[1], Solis Norton[3], Cord Heuer[4], Milan Guatam[4], Jeroen De Buck[5], Christina Ahlstrom[6], Hinrich Vogue[7]

[1]AgResearch, Palmerston North, New Zealand. [2]DM Collins Consulting, Wellington New Zealand. [3]Johne's Management Ltd., Dunedin, New Zealand. [4]Massey University, Palmerston North, New Zealand. [5]University of Calgary, Calgary, Canada. [6]Epi-interactive, Wellington, New Zealand. [7]LIC, Hamilton, New Zealand

11.30 O-02.8 **The relative frequencies of four major strain types of Mycobacterium avium subsp. paratuberculosis in Canadian dairy herds**

Jeroen De Buck[1], Christina Ahlstrom[1], Herman W. Barkema[1]

[1]Department of Production Animal Health, University of Calgary, Faculty of Veterinary Medicine, Canada

11.45 O-02.9 **MOLECULAR EPIDEMIOLOGY OF MAP INFECTION IN CHILE**

Cristobal Verdugo[1], Camilo Tomckoviak[1], Carolina Avilez[1], Miguel Salgado[1]
[1]Instituto de Medicina Preventiva Veterinaria, Universidad Austral de Chile, Chile

12.15-13.15 LUNCH

from 13.15 to 17.00

13.15-14.45 POSTER SESSION

14.45-15.15 TEA BREAK

DIAGNOSTIC AND DETECTION

15.15 **Key note Lecture: Pr S. Mc Kenna**

IMMUNITY & DIAGNOSTIC

16.00 O-03.1 **USING IMMUNOLOGICAL TESTS FOR EARLY IDENTIFICATION OF FARMS WITH MAP EXPOSURE**

Kumudika de Silva[1], Douglas Begg[1], Auriol Purdie[1], Karren Plain[1], Richard Whittington[1]
[1]School of Life and Environmental Sciences, Faculty of Veterinary Science, The University of Sydney, Australia

16.15 O-03.2 **ASSOCIATIONS BETWEEN AVIAN PURIFIED PROTEIN DERIVATIVE HYPERSENSITIVITY AND MAP ANTIBODY ELISA IN IRISH DAIRY COWS.**

Aideen. E. Kennedy[1,2], Noel Byrne[1], Jim O'Mahony[2], Riona. G. Sayers[1]
[1]Animal & Bioscience Research Department, Animal & Grassland Research and Innovation Centre, Teagasc, Moorepark, Fermoy, Co. Cork, Ireland. [2]Department of Biological Sciences, Cork Institute of Technology, Bishopstown, Co. Cork, Ireland

16.30 O-03.3 **IMMUNOLOGICAL ANALYSIS OF 35 RECOMBINANT ANTIGENS OF M. AVIUM SUBSP PARATUBERCULOSIS IN MICE AND CATTLE**

V. Roupie[1#*], S. Holbert[2#], S. Viart[3], C. Tholoniat[4], B. Leroy[3], D. Cappoen[1], M. Branger[2], F. Jurion[1], M. Govaerts[5], C. Van Den Poel[1], B. Lamoureux[4], J-J Letesson[6], L. Malherbe[4], R. Wattiez[3], F. Biet[2], K. Huygen[1]
*[1]Service Immunology, WIV-ISP-Site Ukkel, Brussels, Belgium. *Present address: Unit "Bacterial Zoonoses of livestock", Operational Direction Bacterial Diseases, Veterinary and Agrochemical Research Centre (CODA-CERVA), Groeselenberg, Brussels, Belgium. [2]UMR1282, Infectiologie et Santé Publique, INRA Centre Val de Loire, University François Rabelais of Tours, F-37380 Nouzilly, France. [3]Department of Proteomics and Microbiology, University of Mons - UMONS, Mons, Belgium. [4]Groupement de Défense Sanitaire Région Centre-Val de Loire, F-36018 Châteauroux, France. [5]Department of infectious and parasitic diseases, Faculty of Veterinary Medicine, University of Liège, Belgium. [6]URBM, Facultés Universitaires de Namur, Belgium*
 # equal contributors

16.45 O-03.4 **THE EFFECT OF INOCULATION DOSE ON THE PRECISE NUMBERS OF MYCOBACTERIUM AVIUM SUBSP. PARATUBERCULOSIS SHED IN FECES OVER TIME**

Caroline Corbett[1], Jeroen De Buck[1], Herman Barkema[1]
[1]University of Calgary, Faculty of Veterinary Medicine, Department of Production Animal Health. 3330 Hospital Drive NW, T2N 4N1, Calgary, Alberta, Canada.

WEDNESDAY 22 JUNE 2014

from 8.30 to 13.15

DIAGNOSTIC AND DETECTION

INFORMATIVE VALUE OF DIAGNOSTIC TESTS

8.30 O-03.5 **COMPARATIVE STUDY OF SIX COMMERCIAL ENZYME-LINKED IMMUNOSORBENT ASSAYS FOR BOVINE PARATUBERCULOSIS DIAGNOSTIC**

B. Gentin[1], L. Foures[2], G. Canteneur[3], J. Anderbourg[4], I. Faes[5], E. Sellal[1]
[1]Biosellal, Lyon, France. [2]GDS55, Verdun, France. [3]GDS57, Metz, France. [4]GDS54, Laxou, France. [5]Segilab – LVD55, Bar-le-Duc, France

8.45 O-03.6 **USE OF LIQUID CULTURE WITH IS-900 PCR CONFIRMATION ON FECAL SAMPLES TO ASSESS EXPOSURE OF YOUNG CALVES TO MYCOBACTERIUM AVIUM SUBSP. PARATUBERCULOSIS IN INFECTED HERDS**

de Marchin Emmanuelle[1], Grégoire Fabien[1], Evrard Julien[1], Houtain Jean-Yves[1]
[1]Regional Association for Animal Identification and Health, Ciney, Belgium

9.00 O-03.7 **EVALUATION OF DIAGNOSTIC SENSITIVITY AND SPECIFICITY OF A NEW ELISA TEST ON SERUM AND MILK SAMPLES USING A BAYESIAN APPROACH**

Alvarez J.[1], **Picasso C.**[1], Patnayak D.[1], Goyal S.[1], Wells S.J.[1]
[1]Department of Veterinary Population Medicine and Veterinary Diagnostic Laboratory, University of Minnesota, St. Paul, MN, 55108, USA.

9.15 O-03.8 **COMPARISON OF A PHAGE-PCR ASSAY TO CULTURE AND ELISA FOR THE DETECTION OF MYCOBACTERIUM AVIUM SUBS. PARATUBERCULOSIS IN DAIRY CATTLE**

Gerrard Z.E.[1], Swift B.[3], Davidson R.[2], Hutchings M.[2], Huxley J.[3], and Rees C.[2]
[1]University of Nottingham, School of Biosciences, Sutton Bonington Campus, Leics, LE12 5RD. [2]SRUC, Animal and Veterinary Sciences, Roslin Institute Building, Easter Bush, Midlothian, EH15 9RG. [3]University of Nottingham, School of Veterinary Science and Medicine, Sutton Bonington Campus, Leics, LE12 5RD, United Kingdom

ENVIRONMENTAL DIAGNOSTIC

9.30 O-03.9 **TEMPERATURE EFFECT ON SURVIVABILITY OF MYCOBACTERIUM AVIUM SUBSPECIES PARATUBERCULOSIS IN A TROPICAL ENVIRONMENT**

Orachun Hayakijkosol[1], Robert Hedlefs[1], Elizabeth Kulpa[1], Wayne Hein[2], Jacqueline Picard[1]
[1]College of Public Health, Medical and Veterinary sciences, James Cook University, Townsville, Queensland, Australia. [2]School of Animal and Veterinary Sciences, The University of Adelaide, Adelaide, South Australia, Australia

9.45 O-03.10 **INFORMATIVE VALUE OF ENVIRONMENTAL AND POOLED SAMPLES TO EVALUATE THE WITHIN-HERD PREVALENCE OF MYCOBACTERIUM AVIUM SUBSPECIES PARATUBERCULOSIS IN GOATS FLOCKS**

Esthel Frederic[1], Valentine Grebert[2], Raphael Guatteo[3], Serge Nouzieres[4]
[1]GDS18, Bourges, France. [2]Clinique Vétérinaire de Bracieux, Bracieux, France. [3]UMR Oniris-INRA 1300 BioEpAR, Nantes, France. [4]GDS41, Blois, France

10.00 O-03.11 **EVALUATION OF THE USE OF A POLYMERASE CHAIN REACTION ASSAY ON OVERGROWN ENVIRONMENTAL SAMPLES CULTURED FOR MYCOBACTERIUM AVIUM SSP. PARATUBERCULOSIS**

Juan C. Arango-Sabogal[1], Olivia Labrecque[2], Julie Paré[3], Julie-Hélène Fairbrother[2], Jean-Philippe Roy[1], Vincent Wellemans[1], Gilles Fecteau[1]

[1]Department of Clinical Sciences, Faculty of Veterinary Medicine, Université de Montréal, Saint-Hyacinthe, Québec, Canada. [2]Laboratory of Epidemiological Animal Surveillance of Québec, Ministry of Agriculture, Fisheries and Food of Québec, Saint-Hyacinthe, Québec, Canada.[3]Canadian Food Inspection Agency, Saint-Hyacinthe, Québec, Canada

10.15-10.45 COFFEE BREAK

10.45 POSTER SESSIONS IN PARALLEL WITH WORKSHOP MODELLING

12.15-13.15 LUNCH

from 13.15 to 17.00

ENVIRONMENTAL DIAGNOSTIC

13.15 O-03.12 **A NOVEL RESOLIGHT-BASED REAL-TIME PCR ASSAY WITH POOLED FAECAL SAMPLES FOR THE SCREENING OF JOHNE'S DISEASE IN CATTLE**

Satoko Kawaji[1], Reiko Nagata[1], Yasutaka Minegishi[2], Yumi Saruyama[3], Akiko Mita[4], Yasuyuki Mori[5]

[1] National Institute of Animal Health, [2]Nippon Gene Co., Ltd., [3]Tochigi prefectural Central Livestock Hygiene Service Center, [4]National Livestock Breeding Center, [5]Zen-noh Institute of Animal Health, Japan

13.30 O-03.13 **DETECTION OF MYCOBACTERIUM AVIUM SUBSPECIES PARATUBERCULOSIS (MAP) IN POWDERED INFANT FORMULA**

Kamal Raj Acharya[1], Plain K. M.[1], Whittington R. J.[1], Dhand N. K[1]

[1]Farm Animal & Veterinary Public Health, Faculty of Veterinary Science, School of Life and Environmental Sciences, The University of Sydney, Sydney, New South Wales 2570, Australia

NEW APPROACHES

13.45 O-03.14 **CIRCULATING MIRNA IN BOVINE SERUM AND THEIR POTENTIAL USE AS NOVEL BIO MARKERS OF EARLY MYCOBACTERIUM AVIUM SUBSP PARATUBERCULOSIS INFECTION**

Ronan G. Shaughnessy[*1], Damien Farrell[*1], Karel Riepema[6], Louise Britton[1], David E. MacHugh[2,5], Bryan Markey[1], Douwe Bakker[6], Stephen V. Gordon[1,3,4,5]

[1]School of Veterinary Medicine, [2]School of Agriculture and Food Science, [3]School of Medicine, [4]School of Bio-molecular and Biomedical Science, [5]Conway Institute, University College Dublin, Dublin, Ireland. [6]Department of Bacteriology and TSEs, Central Veterinary Institute of Wageningen University, Edelhertweg 15, 8200 AB Lelystad, The Netherlands.

** Equal author contribution*

14.00 O-03.15 **DIAGNOSTIC POTENTIAL OF A PEPTIDE-MEDIATED MAGNETIC SEPARATION (PMS)-PHAGE ASSAY APPLIED TO MILK RELATIVE TO FAECAL- AND BLOOD-BASED TESTS**

Irene Grant[1], Lorna O'Brien[1]

[1] Institute for Global Food Security, Queen'sUniversity Belfast, Northern IrelandSam Strain, Animal Health and Welfare Northern Ireland, Armagh, Northern Ireland

14.15 O-03.16 **SPECIFIC AND SENSITIVE DETECTION OF VIABLE MAP WITHIN 6 H USING BACTERIO PHAGE**

Ben Swift[1], Le Van Hung Vuong [1], Catherine Rees[1]
[1] *University of Nottingham, School of Biosciences, UK*

14.30 O-03.17 **RECOMBINANT SECRETORY PROTEINS BASED DIAGNOSIS FOR DETECTION OF DIFFERENTIAL IMMUNE RESPONSE TO MYCOBACTERIUM AVIUM SUBSPECIES PARATUBERCULOSIS INFECTION AND VACCINATION IN DOMESTIC LIVESTOCK**

Kundan Kumar Chaubey[1], Saurabh Gupta[1], Jagdip Singh Sohal[2], Sujata Jayaraman[2], Bjorn John Stephan[1], Manju Singh[1], Mukta Jain[2], Naveen Kumar[3], **Shoor Vir Singh**[1]
[1]*Microbiology laboratory, Animal Health Division, Central Institute for Research on Goats, Makhdoom, PO-Farah, Mathura, 281122, India.* [2]*Amity Institute of Microbial Technology, Amity University Rajasthan, Kant Kalwar, NH 11C Delhi-Jaipur Highway, Jaipur, 303 002, Rajasthan.*[3]*Veterinary Type Culture Collection, National Research Centre on Equines, Hisar, Haryana-125001, India*

14.45-15.15 TEA BREAK

EXPOSURE AND TRANSMISSION

15.15 **Key note Lecture: Pr C. Heuer**

16.00 O-04.1 **TRANSMISSION OF INFECTION AMONG GROUP-HOUSED CALVES INOCULATED WITH MYCOBACTERIUM AVIUM SUBSP. PARATUBERCULOSIS**

Caroline Corbett[1], Jeroen De Buck[1], Karin Orsel[1],Herman Barkema[1]
[1]*University of Calgary, Faculty of Veterinary Medicine, Department of Production Animal Health. 3330 Hospital Drive NW, T2N 4N1, Calgary, Alberta, Canada*

16.15 O-04.2 **PREVALENCE OF MYCOBACTERIUM AVIUM SUBSP. PARATUBERCULOSIS (MAP) IN PASTEURISED MILK AND EVIDENCE FOR MECHANISMS OF SURVIVAL**

A. Saeedi[1], Gerrard Z.[1], Swift B.[2], and Rees C.[1]
[1]*University of Nottingham, School of Biosciences, Sutton Bonington Campus, Leics, LE12 5RD.* [2]*University of Nottingham, School of Veterinary Science and Medicine, Sutton Bonington Campus, Leics, LE12 5RD, UK*

16.30 O-04.3 **REGIONAL DIFFERENCES IN JOHNE'S DISEASE PREVALENCE BASED ON ENVIRONMENTAL CULTURE AND BULK MILK TESTING**

D. Kelton[1], H. Barkema[2], C. Bauman[1], C. Pickel[2]
[1]*University of Guelph, Guelph, Ontario, Canada* [2]*University of Calgary, Calgary, Alberta, Canada*

16.45 O-04.4 **TRUE WITHIN HERD PREVALENCE DISTRIBUTION AND RISK FACTORS TO MAP INFECTION AMONG DAIRY HERDS IN CHILE**

Cristobal Verdugo[1], Maria Francisca Valdez[1], Miguel Salgado[1]
[1]*Instituto de Medicina Preventiva Veterinaria, Universidad Austral de Chile, Chile*

17.00-18.00 IAP GENERAL MEETING

THURSDAY 23 JUNE 2014

from 8.30 to 13.45

ONE HEALTH & CONTROL PLANS

8.30 **Key note Lecture : L. Waddell**

MODELLING

9.15 O-05.1 **MODELING PARATUBERCULOSIS IN A TYPICAL SHEEP FLOCK IN NEW ZEALAND**

Marquetoux N.[1], Mitchell R.[2], Wilson P.[1], Stevenson M.[3], Ridler A.[1], Heuer C.a[1]
[1]EpiCentre, Institute of Veterinary, Animal and Biomedical Sciences,Massey University,New Zealand. [2]Department of Population Medicine and Diagnostic Sciences, College of Veterinary Medicine, Cornell University,USA. [3]Faculty of Veterinary and Agricultural Sciences, The University of Melbourne, Parkville Victoria 3052 Australia

9.30 O-05.2 **SPREAD AND CONTROL OF BOVINE PARATUBERCULOSIS IN AN ENZOOTIC CATTLE REGION: A MULTI-SCALE MODEL TO EVALUATE COMPLEX STRATEGIES COMBINING BIOSECURITY AND TEST-AT-PURCHASE**

Gaël Beaunée[1,2], Elisabeta Vergu[2], Alain Joly[3], **Pauline Ezanno**[1]
[1]INRA, ONIRIS, LUNAM Université, UMR1300 BioEpAR, CS40706, F-44307 Nantes, France. [2]INRA, UR1404 MaIAGE, F-78350 Jouy-en-Josas, France. [3]Groupement de Défense Sanitaire de Bretagne, 56019 Vannes, France

VACCINATION

9.45 O-05.3 **CHANGES IN PREVALENCE OF OVINE PARATUBERCULOSIS FOLLOWING EXTENDED VACCINATION WITH GUDAIR OVER MORE THAN A DECADE**

Navneet Dhand[1], Peter A. Windsor[1], Richard J Whittington[1]
[1]The University of Sydney, Australia

10.00 O-05.4 **CHANGES IN PREVALENCE OF OVINE PARATUBERCULOSIS IN SPECIFIC COHORTS OF VACCINATES OVER FIVE, TWO-YEAR INTERVALS**

Jeffrey Eppleston[1], Navneet Dhand[1], Peter A. Windsor[1], Richard J Whittington[1]
[1]The University of Sydney, Australia

10-15 O-05.5 **ECONOMIC EFFECTS OF JOHNES DISEASE AND ESTIMATES OF VACCINE EFFICACY ON SHEEP FARMS IN NEW ZEALAND**

Milan Gautam[1], Peter Anderson[2], Anne Ridler[1], Cord Heuer[3]
[1]Institute of Veterinary, Animal and Biomedical Sciences, Massey University. [2]The Vet Centre Marlborough, Blenheim. [3]EpiCentre, Institute of Veterinary, Animal and Biomedical Sciences, Massey University, New Zealand

10-30 O-05.6 **NEW CRITERIA TO DIFFERENTIATE PARATUBERCULOSIS VACCINATION FROM BOVINE TUBERCULOSIS INFECTION IN THE CERVICAL COMPARATIVE SKIN TEST**

Serrano M.[1], Elguezabal N.[1],Urkitza A.[2], Geijo M.V.[1], Molina E.[1], Arrazuria R.[1], A. Sevilla I.[1], Vordermeier M.[3], Whelan P.[3], Juste R.A.[1,4], Garrido J.M[1]
[1]NEIKER - Tecnalia, Basque Institute for Agricultural Research and Development, Derio, Bizkaia, Spain. [2]Veterinary Surgeon, Gernika, Bizkaia, Spain. [3]Department of Bovine Tuberculosis. Ahvla, Surrey, United Kingdom. [4]SERIDA, Department of Agriculture of the Regional Government of the Principality of Asturias, Deva, Asturias, Spain

10-45-11.15 COFFEE BREAK

CONTROL PLANS

11.15 O-05.7 **The Thuringian bovine Johne's disease control program – results and conclusions for the future control strategy**

Karsten Donat[1], Heike Köhler[2]
[1] ThueringerTierseuchenkasse, Animal Health Service, Jena, Germany. [2]Institute of Molecular Pathogenesis, Friedrich-Loeffler-Institut, Federal Research Institute for Animal Health, Jena, Germany

11.30 O-05.8 **Milk quality assurance programme for paratuberculosis: decreasing proportion of ELISA-positive heifers**

Weber MF[1], Heuer C[2], Aalberts M[1], Schukken YH[1]
[1]GD Animal Health, Deventer, the Netherlands. [2]EpiCentre, Massey University, Palmerston North, New Zealand

11.45 O-05.9 **Implementation of control measures for bovine paratuberculosis in Swiss dairy and beef herds**

Mireille Meylan[1], Myriam Anderegg[1]
[1]Clinic for Ruminants, Vetsuisse Faculty, University of Berne, Switzerland

12.00 O-05.10 **Shifting strategy for paratuberculosis management in Queensland**

Lawrence Gavey[1]
[1]Principal Veterinary Officer (Animal Disease Containment), Biosecurity Queensland, Department of Agriculture and Fisheries, PO Box 102, Toowoomba, Queensland, Australia

12.15 O-05.11 **Success and failure in the control of Paratuberculosis in dairy herds by risk management**

Richard Sibley[1], Peter Orpin[2]
[1] BVSc HonFRCVS, Westridge Veterinary Group, Witheridge, UK. [2]BVSc MRCVS Park Vet Group, Leicester, UK

12.30 O-05.12 **Are we feeding live Mycobacterium avium subsp. paratuberculosis in calf milk replacer?**

James C. Tarrant[1], Brenna Kunkel[1], Sheila McGuirk[1], Antonio C.G. Foddai[2], Irene R. Grant[2], Michael T. Collins[1]
[1]School of Veterinary Medicine, University of Wisconsin-Madison, Wisconsin, USA. [2]Institute for Global Food Security, School of Biological Sciences, Queen's University Belfast, Northern Ireland

12.45-13.45 Lunch

from 13.45 to 17.00

13.45 Poster Session (Exposure & Transmission / One Health & Control plans)

14.45-15.15 Tea break

FARMERS PERCEPTION & ATTITUDES

15.15 O-05.13 **Johne's disease management in UK dairy herds participating in a quarterly testing programme - results of a farmer survey**

Karen Bond[1]
[1]National Milk Records, Royal Veterinary College, Carnforth, UK

15.30 O-05.14 **MOTIVATING ON-FARM CHANGE FOR JOHNE'S DISEASE CONTROL USING PEER LEARNING AND WHITEBOARD VIDEOS**

Steven Roche[1], Dr. David Kelton[1], Dr. Ann Godkin[2]

[1]Department of Population Medicine, Ontario Veterinary College, University of Guelph. [2]Ontario Ministry of Agriculture, Food and Rural Affairs, Canada

15.45 O-05.15 **THE USE OF A WEB BASED PREVALENCE PREDICTOR AND RISK MODEL TO ENGAGE FARMERS IN JOHNE'S CONTROL**

Peter Orpin[1,2], Richard J. Sibley[3]

[1]Park Vet Group, Whetstone, Leicestershire. [2]Special Lecturer, Nottingham University, Nottingham, UK. [3]Westridge Veterinary Group, Witheridge, UK

16.00 O-05.16 **DAIRY FARMERS' PERCEPTIONS TOWARDS IMPROVING ON-FARM JOHNE'S DISEASE PREVENTION AND CONTROL**

Caroline Ritter[1], Jolanda Jansen[2], Keliesha Roth[1], John P. Kastelic[1], Cindy L. Adams[3], Herman W. Barkema[1]

[1]Department of Production Animal Health, Faculty of Veterinary Medicine, University of Calgary, Calgary, AB, Canada, T2N 4N1. [2]St. Anna Advies, 6525 Nijmegen, The Netherlands. [3]Department of Veterinary Clinical and Diagnostic Sciences, Faculty of Veterinary Medicine, University of Calgary, Calgary, AB, Canada, T2N 4N1

16.15-17.00 **SYNOPSIS ON THE CONFERENCE AND CLOSURE**

19.00-23.30 CONFERENCE DINNER IN THE DUCHESSE ANNE CASTLE

NOTES

NOTES

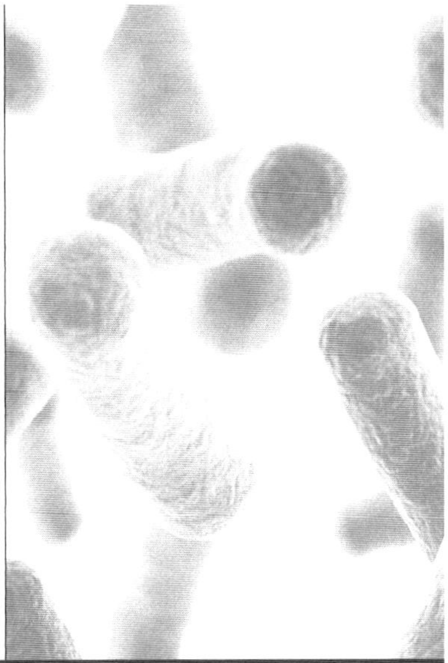

SESSION 1

Infection course

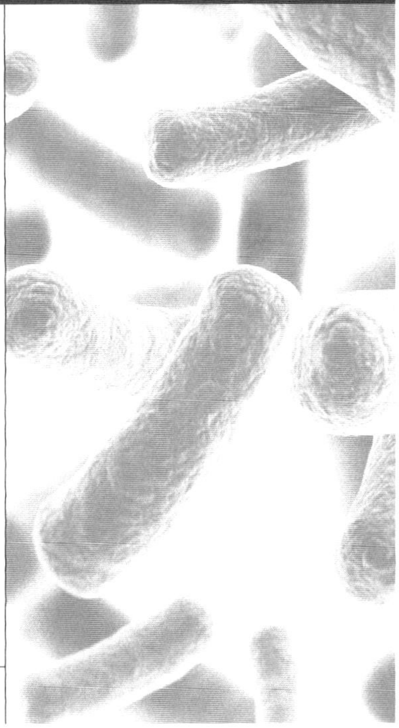

PROTECTIVE IMMUNITY: CURRENT CHALLENGES IN TB VACCINE DEVELOPMENT

Rasmus Mortensen, Statens Serum Institut, Department of Infectious Disease Immunology, Denmark.

ORAL

O-01.1
IN VITRO EXPERIMENTAL WORK AND AGENT-BASED MODELING OF GRANULOMA FORMATION CAUSED BY MYCOBACTERIUM AVIUM SUBSP. PARATUBERCULOSIS

Shigetoshi Eda[1], J. Hunter Rice[1], Alyson Marek[1], Sarah M. Alouani[1], and Judy Day[2]

[1]*Department of Forestry, Wildlife and Fisheries, University of Tennessee Institute of Agriculture, USA*
[2]*Department of Mathematics, University of Tennessee Knoxville, USA*

Mycobacterium avium subsp. paratuberculosis or MAP is the causative agent of bovine Johnes Disease (JD), a potentially debilitating illness in cattle that can have significant economic impacts particularly in the dairy industry. In the U. S. there is currently only one approved vaccine, known as Mycopar, for the control of JD in cattle; however, Mycopar was shown to be only partially effective as it reduces fecal shedding and clinical symptoms but does not prevent new infections. More efficient vaccines will likely be developed when there is a better understanding of the protective host immune responses in MAP-infected animals.The initial and primary battleground for an animal's immune system to attack MAP occurs in intestinal tissues within specifically formed structures called granulomas. A proportion of MAP-infected animals could control the infection under a dominant cellular immune response. However, detailed mechanisms of bacterial killing in JD granuloma are largely unknown. We developed an in vitro granuloma assay to generate new data on dynamics of granuloma formation, macrophage/MAP viability, and cytokine expression.Bovine peripheral blood mononuclear cells were separated into non-adherent (mostly lymphocytes) and adherent (macrophages) cells. Macrophages were infected with MAP and, after removing the extracellular MAP, supernatant of MAP-exposed lymphocytes was added to the culture and incubated for up to 2 weeks. Granuloma-like cell clusters (GLCCs) were formed specifically in MAP-infected culture and contained multinucleated cells. The number of GLCCs increased over time and peaked at 7-8 days after infection. Macrophage viability decreased over time but MAP infected macrophages survived longer than uninfected macrophages. Motility of MAP-infected macrophages was much lower than that of uninfected macrophages.
The number of MAP in macrophages slightly increased over a 7-day culture with no significant change in viability. IFN-γ, IL-1β, IL-4, and IL-10 were induced by MAP infection at 4 day post-infection.We are currently developing an agent-based model incorporating the data collected from the in vitro granuloma model. With the model, we aim to better understand the contributions of several exogenous factors (e.g. cytokines) that contribute to the dynamics of granuloma formation and killing of MAP.

Keywords :
Mycobacterium avium subsp. paratuberculosis, macrophages, granuloma, agent-based model

O-01.2
EARLY INTERACTIONS BETWEEN BOVINE MONOCYTE DERIVED MACROPHAGES AND MYCOBACTERIUM AVIUM SUBSP. PARATUBERCULOSIS

Heather Mathie[1], Joanne Stevens[1], Elizabeth Glass[1], Jayne Hope[1]

[1]*The Roslin Institute, The University of Edinburgh, Edinburgh, Scotland*

Mycobacterium avium subsp. paratuberculosis (MAP) infection of cattle has a damaging economic impact and is a cause for concern regarding potential zoonotic transmission. There are significant difficulties regarding preventing, detecting, and eradicating MAP infection in cattle herds, many of which stem from the ability of MAP to evade the host immune response (and thus diagnosis). This evasion is associated with the ability of MAP to persist within macrophages for prolonged periods of time. The precise mechanisms of this host-pathogen interaction are not fully understood; we hypothesise that the earliest interactions between MAP and host macrophages, including the mechanism of uptake, may play a pivotal role in determining the outcome of infection. In order to investigate this, we infected bovine monocyte derived macrophages (MDM) with two strains of MAP (the well-established K10 strain and strain C49, a recent clinical isolate from cattle) in order to model the infection in vitro. Following infection of MDM with the C49 strain, numbers remained fairly constant over a 24 hour time-course, whereas the K10 strain was completely eradicated by 24 hours post infection; this difference in survival is indicative of a loss of virulence in the K10 strain through lab adaptation. Due to the difference observed between the two strains in their ability to survive intracellularly, the early immune response of MDMs to each strain was analysed, shedding light on mechanisms which could be important for early clearance. By assessing cell surface molecule expression, gene expression, cytokine secretion, nitrite and reactive oxygen species production and phagosome acidification, a detailed picture of the early macrophage response was formed for both strains. We also demonstrated that the presence of serum antibody responses to MAP can impact uptake and survival of MAP within macrophages and impact their downstream effector functions. This suggests that the mechanism of uptake, for example via Fc or complement receptors, could play a role in determining downstream responses to MAP infection.

Keywords :
Macrophage, Innate response, MAP-host interaction

O-01.3

Rabbit gut microbiota modifications induced by Mycobacterium avium subspecies paratuberculosis infection and diet changes

Rakel Arrazuria[1], Elena Molina[1], Iker A.Sevilla[1], Miriam Serrano[1], Hooman Derakhshani[2], Ehsan Khafipour[2,3], Valentín Perez[4], Joseba Garrido[1], Ramon A. Juste[1,5], Natalia Elguezabal[1]

[1]NEIKER-Tecnalia, Basque Institute for Agriculture Research and Development, Derio, Bizkaia, Spain; [2]Department of Animal Science, University of Manitoba, Winnipeg, MB, Canada; [3]Department of Medical Microbiology, University of Manitoba, Winnipeg, MB, Canada; [4]Department of Animal Health, Faculty of Veterinary Medicine, University of León, Spain; [5]SERIDA, Department of Agriculture of the Regional Government of the Principality of Asturias, Deva, Asturias, Spain

MAP infection has been associated with dysbiosis of intestinal microbiota in ruminants. In addition, diet is one of the key factors affecting the balance of microbial populations in the digestive tract and it has shown to modulate MAP infection in a rabbit model. In this study, we intended to further analyze this alteration in infection progression associated to diet changes and see if variations in microbial composition were occurring in a MAP rabbit infection model under different dietary conditions.

The infectious status and microbial composition was studied in 20 New Zealand rabbits. Half of those rabbits were fed a regular diet during the whole experiment (challenged n=5, control n=5). The remaining rabbits were fed a high fiber diet during the challenge (challenged n=5, control n=5). The infection was assessed by qPCR and histopathology of gut associated lymphoid tissues. The microbiota of sacculus rotundus and cecal content was studied by 16S rRNA gene sequencing. QIIME software package was used for microbial community analysis. Predictive functional profiling based on KEGG metabolic pathway was performed using PICRUSt.

All infected animals fed a high fiber diet and four of five (80%) fed a regular diet showed a positive result for qPCR in at least one tissue. However, infected animals fed a regular diet showed more severe histopathological lesions compatible with MAP infection than infected animals fed a high fiber diet (mean of lesions severity index 0.51 and 0.17 respectively). Analysis of the microbial community showed some bacteria repeatedly positively associated with infection (families Dehalobacteriaceae, Coriobacteriaceae and Mogibacteriaceae) and repeatedly negatively associated with infection (genera Anaerostipes and Coprobacillus), in different samples under different diets. However, other bacteria (Enterobacteriaceae family and ML615J-28 order) were positively associated with infection in some circumstances and negatively in others. Microbial predicted functions showed the enrichment associated with cellular processes in sacculus rotundus of infected animals and the genetic information processing in cecal content of control animals.

In conclusion, infected animals fed a regular diet showed more severe histopathological lesions whereas changes in the gut microbiota involving microbial functions were observed with both diets.

Keywords :
Mycobacterium avium subsp. paratuberculosis; rabbit; animal model; dietary changes, metagenomics; gut microbiota.

O-01.4
QUANTIFYING RESILIENCE TO ARTIFICIAL SYSTEMIC MAP INFECTION USING HISTOPATHOLOGY AND THE IDENTIFICATION OF HEPATIC EPITHELIOIDE MACROPHAGE (HEM) MICRO-GRANULO MAS FROM SERIAL LIVER BIOPSIES IN SHEEP.

Stefan Lindsay Smith[1], Professor Peter Wilson[1], Professor Dave West[1], Professor Cord Heuer[2], Associate Professor Paul Chambers[1]

[1] *Institute of Veterinary, Animal and Biomedical Sciences, Massey University, Palmerston North, New Zealand;* [2] *The Epi Centre, Massey University, palmerston North, New Zealand*

Production losses from Map infection in sheep have been difficult to quantify with recorded losses attributable only to clinical disease. Systemic Map infection has been recorded in sheep affected by clinical Johne's disease(JD) yet not in healthy sheep suggesting it may have a role to play in the establishment of JD and contribute to these production losses. The presence of Map specific HEM from liver biopsy histopathology provides, as a surrogate measure, a method to identify the occurrence of systemic Map infection. Currently HEM have only been identified in sheep with clinical JD suggesting a shared prevalence(0.13%). The aims of this longitudinal study were to determine the prevalence of HEM formation in artificially challenged lambs, the time taken for their formation and whether resilience to systemic Map infection occurs, the cost of this resilience and the relationship between systemic Map infection and clinical JD in sheep. Thirty six, three month old, Map free lambs were randomly selected and separated into two equal groups of 18 animals. Group one(G1) lambs were artificially orally dosed, on 10 different occasions, 3 days apart with >1x109 Map organisms using a homogenate of terminal ileum and mesenteric lymph nodes sourced from a ewe diagnosed with clinical JD. Group two(G2) lambs were orally dosed on the same occasions with 10mL sterile phosphate buffered saline. All lambs were pasture grazed outdoors for 820 days with the groups separated by a fence to prevent cross contamination. Serial liver biopsies were collected throughout with HEM formation recorded from t=51-114 days in all of G1 and none of G2. Formation of HEM preceded sero-conversion with titre optical densities increasing proportionally to HEM prevalence. Four lambs from G1 developed terminal JD between t=290-450 days with 14/18 making full recoveries. This resilience to systemic Map infection appears to occur with an inverse relationship recorded between HEM prevalence and growth rates/live weights, with G1 having smaller skeletal structures and 11kg liveweight differences at t=820 days. Whether similar resilience occurs in naturally infected sheep has yet to be quantified suggesting care needs to be taken when comparing artificial challenge models with natural infection.

Keywords :
Mycobacterium avium subsp paratuberculosis, Map, Hepatic, micro-granuloma, HEM, histopathology, resilience, systemic, artificial, production

O-01.5
EXPERIMENTAL INFECTION OF CATTLE WITH MYCOBACTERIUM AVIUM SUBSP. PARATUBERCULOSIS: RESULTS FROM THE FIRST TWO YEARS

Louise Britton[1], Colm Brady[2], Joseph Cassidy[1], Jim O'Donovan[3], Nola Leonard[1], Eamon Gormley[1], Ronan Shaughnessy[1], Stephen V. Gordon[1], Bryan Markey[1]

[1]*School of Veterinary Medicine, University College Dublin, Ireland.* [2]*Department of Agriculture, Food and the Marine, Central Veterinary Research Laboratory, Backweston, Kildare, Ireland.* [3]*Department of Agriculture, Food and the Marine, Regional Veterinary Laboratory, Model Farm Road, Cork, Ireland*

A long-term Mycobacterium avium subspecies paratuberculosis (MAP) bovine experimental infection study has been established as part of the IconMAP project, funded by the Irish Department of Agriculture, Food and the Marine (DAFM). Its aim is to facilitate a search for novel biomarkers of infection status by creating a biobank of tissue and blood samples from animals of known status for test evaluation and validation. The calves in the infected group (n=35) received 3.8X109 CFU MAP twice orally at approximately 4 weeks of age, whereas the control calves (n=20) received a placebo. Their MAP infection status is being monitored using currently available diagnostic tests. The IFN-γ assay (Bovigam®) was used to assess the cell mediated immune response against MAP infection; results to date indicate that the experimental infection was successful. Faecal samples were cultured for MAP using the TREK ESP® para-JEM® system; samples from the pre-infection time-point and months 3, 6, 9, 12, 16, 20 and 24 post-infection (PI) have been cultured and are negative. The humoral immune response has been evaluated using a commercial ELISA (Idexx); serological results are available up to 24 months PI - a single animal tested positive at month 10 PI, two animals tested positive at month 16 PI, six animals tested positive at month 20 PI and four animals tested positive at month 24 PI. At the end of the first and second years PI, thirteen animals (n=8 infected; 5 controls) were euthanized for detailed post-mortem examination. At month 12 PI, only minor pathological differences were seen between the groups and ileal tissue samples were MAP culture negative. At month 24 PI, one infected animal showed thickening and corrugation within the ileum and six infected animals showed enlarged ileocaecal lymph nodes suggestive of MAP infection; histopathological and tissue culture results are pending. These results clearly demonstrate the difficulties of identifying subclinically infected cattle and emphasize the requirement for novel diagnostic biomarkers of infection status.

Keywords :
IconMAP, experimental infection, diagnostics, biobank, biomarkers

O-01.6

IDENTIFICATION AND EVALUATION OF PROGNOSTIC BIOMARKERS FOR MYCOBACTERIUM AVIUM SUBSP. PARATUBERCULOSIS INFECTION

Hong-Tae Park[1], Hyun-Eui Park[1], **Han Sang Yoo**[1,2]

[1]*Department of Infectious Diseases, College of Veterinary Medicine, Seoul National University, Seoul, 08826, Republic of Korea.* [2]*Institute of Green-Bio Science and Technology, Seoul National University, Pyeongchang, 25354, Republic of Korea*

Mycobacterium avium subsp. paratuberculosis (MAP) is the causative agent of Johne's disease, a chronic debilitating disease affecting ruminants worldwide. MAP is slow-growing bacterium that has very long latent periods. To control the disease, a new diagnostic method has been required to identify the early stage of infection due to the excretion of MAP in feces of cattle at sub-clinical stage. Based on this situation, attempts were made to identify biomarkers that show early responses to MAP infection. Host transcriptional responses of mouse model were analyzed in mouse system using a microarray. Subsequently, gene expression changes were analyzed in naturally infected cattle that were grouped based on ELISA and fecal PCR results (Test 1: ELISA(-)/PCR(+); Test 2: ELISA(+)/ PCR(-); Test 3: ELISA(+)/PCR(+)).
MAP-infected mice at 3 and 6 weeks p.i. showed up-regulation of initial innate immune reaction related genes. Concurrently, they were also suggested to express M2 macrophage phenotypes and canonical pathways related to the T cell responses. Through the functional analysis of transcriptional profiles on cattle, the results showed down-regulated production and metabolism of reactive oxygen species in the Test1 group, activation of pathways related to the host-defense response against MAP (LXR/RXR activation and complement system) in the Test2 and Test3 groups, and anti-inflammatory response (activation of IL-10 signaling pathway) only in the Test3 group.
From the data of gene expression changes in mouse and cattle, total of 23 potential biomarker candidates were selected and they were validated by naturally infected cattle using the real-time PCR assay. In sub-clinically infected animals, 17 genes were up-regulated. Of the up-regulated genes, six genes (CXCR3, HP, HGF, LTF, TFRC, and GBP6) showed significant differences between experimental groups. The results suggest that 10 genes (KLRB1, S100A8, S100A9, LTF, HGF, HP, CXCR3, SERPINE1, GBP6, and TFRC) played essential roles in immune response to MAP during the subclinical stage and might be therefore be useful as prognostic biomarkers.
This work was carried out with the support of «Cooperative Research Program for Agriculture Science & Technology Development (PJ008970)» RDA and BK21 PLUS Program, Republic of Korea.

Keywords :
Mycobactrerium avium subsp paratuberculosis, Map, Hepatic, micro-granuloma, HEM, histopathology, resilience, systemic, artificial, production

O-01.7

DIFFERENTIAL SYSTEMIC AND LOCAL CYTOKINE-RESPONSES AFTER EXPERIMENTAL INFECTION OF GOATS WITH MYCOBACTERIUM AVIUM SUBSP. PARATUBERCULOSIS (MAP) OR MYCOBACTERIUM AVIUM SUBSP. HOMINISSUIS (MAH)

Heike Köhler[1], Anja Zigan[1], Jan Schinköthe[1], Petra Möbius[1], Elisabeth M. Liebler-Tenorio[1]

[1]*Institute of Molecular Pathogenesis, Friedrich-Loeffler-Institut, Jena, Germany*

The kinetics and intensity of systemic and local cytokine responses to MAP and MAH were compared in an experimental goat model.
21 goat kids each were orally inoculated with MAP or MAH, receiving total doses of 6,3x108 or 2,1x1010 cfu per animal, respectively. Clinical symptoms and course of specific interferon-gamma (IFN-γ) responses were monitored regularly up to 48 weeks post inoculation (wpi) in inoculated animals and 10 age-matched controls. Gene expression of the cytokines IFN-γ, IL-1β, IL-10, IL-12p40, IL-17, IL-18, IL-23, TGF-β and TNF-α was examined in organized gut associated lymphatic tissue (oGALT) and intestinal lymph nodes (ILN) when goats were necropsied at 3-12 wpi, 24 wpi and 48 wpi.
MAP-inoculated and control goats were clinically healthy throughout the experiment until necropsy at 48 wpi. The MAH-inoculated goats developed mild transient symptoms up to progressive disease (fever, depression, weight loss). 9 of them died or were euthanized 3-10 wpi. Already 7-10 wpi, a strong antigen-induced IFN-γ response was noted in MAP- and MAH-inoculated animals. While the initial response was massive in the MAH-inoculated goats, it only gradually increased up to 15-18 wpi in the MAP-inoculated animals. After MAP infection, gene expression of pro- and anti-inflammatory cytokines did not differ markedly over time, but showed a large individual variation. As an exception, IL-17 was down-regulated in oGALT of animals necropsied 48 wpi in comparison to 12 and 24 wpi. The majority of the examined cytokines were up-regulated in oGALT of MAH-inoculated goats at 3-10 wpi in contrast to those necropsied at 48 wpi. Gene expression of IL-1β, IL-17 and TGF-β differed significantly. Individual variation was less pronounced.
IL-23 expression was increased in oGALT and ILN of MAP-inoculated goats at 48 wpi compared to controls.
It can be concluded, that IL-17, TNF-α, IL-1β, and TGF-β contribute to the differences in tissue damage after infection with MAP versus MAH. The magnitude of IL-23 expression may be related to the extent of bacterial organ burden.

Keywords :
Gene expression, real-time RT-PCR, gut-associated lymphatic tissue, intestinal lymph nodes

O-01.8
The importance of cholesterol in MAP infection of ruminants

Matt Johansen[1], de Silva K[1], Plain KM[1], Begg D[1], Biet F[2], Whittington RJ[1], Purdie AC[1].

[1]*The University of Sydney, Camden, Australia.* [2]*Institut National de la Recherche Agronomique, Nouzilly, France*

Cholesterol plays an important role in the establishment of mycobacterial infections. Mycobacterial species such as Mycoacterium tuberculosis are capable of utilising cholesterol as a primary energy source in culture. M. leprae localises to cholesterol-rich areas of an infected cell, which holds significance for the intracellular niches created by mycobacteria. Despite the implications for other mycobacterial species, there has been very little research examining the relationship between cholesterol and Mycobacterium avium subsp. paratuberculosis (MAP). This study examined the role of both serum and intracellular cholesterol during the early stages of MAP infection in both sheep and cattle in vivo and in vitro. Using a well-established infection model, sheep and cattle were exposed to MAP and blood and faecal samples were collected at monthly intervals. Total serum cholesterol changed significantly in exposed animals during the first few months of infection in both sheep and cattle when compared to the control cohorts. In in vitro infection experiments using monocytes from MAP non-exposed cattle and sheep and fluorescent microscopic techniques demonstrated that GFP-tagged MAP co-localised to cholesterol-rich domains within the cell. In addition, changes in the expression of a number of cholesterol-associated genes within the macrophage suggest that MAP is capable of altering cholesterol metabolism of the infected cell. Thus previously unexplored mechanisms within the intracellular environment created by MAP during the early stages of infection may be important for understanding the survival and persistence of MAP.

Keywords :
Cholesterol, pathogenesis

O-01.9

Case definitions for paratuberculosis – the need for standards

Richard Whittington[1], Kumudika de Silva[1], Douglas Begg[1], Auriol Purdie[1], Navneet Dhand[1], **Karren Plain**[1]

[1]*School of Life and Environmental Sciences, Faculty of Veterinary Science, The University of Sydney, Australia*

A case definition is a set of uniform criteria used to classify animals or groups of animals into disease or other outcome categories. Counts generated from a case definition can be used to estimate prevalence, incidence, or other measures of disease frequency. The outcomes can be used to conduct logistic regression or survival analyses to evaluate their association with husbandry, immunological, gene expression or other variables. Agreed definitions provide for rigor in analysis, and enable meaningful comparisons between experiments, laboratories or data from different research teams in different countries.
Johne's disease presents as a range of complex phenotypes, and the chronicity of disease pathogenesis contributes to difficulties with disease characterization at each stage of progression. It is this variation, often combined with a lack of rigor in case definition, that has led to real or apparent lack of agreement between the results of different studies, be they focused on simple validation of diagnostic tests, vaccines or complex genome wide associations.
Johne's disease is a comprehensive term used to describe all forms of infection with MAP and the resulting disease. It does not necessarily imply that the animal has outward signs of disease. We have developed and applied rigorous case definitions using descriptors that were applied as outcome variables in many studies: exposed, diseased (clinical or sub-clinical), infected (clinical or subclinical), infectious, recovered or resistant. These definitions were based on weight loss, MAP shedding, immunological responses, and the presence of gross and/or histopathological lesions and MAP (culture and/or PCR) in intestinal tissues. We describe special circumstances where variation of the criteria is possible based on objective information, usually from experimental infection trials.In this paper we present the rationale and definitions applied and seek international dialogue on their adequacy, applicability and adoption.

Keywords :
Case definition, paratuberculosis, standards

O-01.10

Effects of vaccination before or after M. a. subsp paratuberculosis (Map) experimental infection in goats

Royo M[1], Fuertes M[1], Fernández M[1], Sevilla IA[2], Arrazuria R[2], Castaño P[1], Ferreras MC[1], Benavides J[1], Elguezabal N[2], Pérez V[1]

[1]*Dpt Sanidad Animal, Instituto de Ganadería de Montaña (CSIC-ULE), Facultad de Veterinaria, Universidad de León, León, Spain.* [2]*Dpt of Animal Health, NEIKER-Instituto Vasco de Investigación y Desarrollo Agrario. Derio, Bizkaia, Spain*

Previous field studies have pointed out that vaccination of adult animals, presumably already infected with Map, decreases the appearance of new clinical cases, suggesting a therapeutic effect. With the aim of investigating the effects of vaccination, a total of 35, 1.5 month-old goat kids were employed. Eight of them were vaccinated subcutaneously with 1ml of an inactivated vaccine (Silirum®), and one month later orally challenged with 1010 CFU of the Map K-10 strain. At the same time, 14 non-vaccinated kids were similarly infected. The rest of the animals were kept as vaccinated (5) or non vaccinated non infected (8) controls. The peripheral immune response was assessed by indirect ELISA and IFN-γ release tests. At 150 days post-infection (dpi), 5 infected and 3 infected and vaccinated kids were pathologically examined. After lesion development verification, five already infected kids were vaccinated with the same product at 180 dpi. At 360 dpi all the animals were slaughtered. A significant increase in IFN-γ production was detected in all the vaccinated animals 1 month post-vaccination, that was significantly lower in those vaccinated after the infection. At 150 dpi, granulomatous lesions were identified both in infected and in vaccinated-infected kids, but the number of granulomas were significantly higher in the former group. Moreover, in vaccinated-infected animals lesions were restricted to the lymphoid tissue (focal forms) and demarcated whereas in the unvaccinated group they were seen in the mucosa either related or not to the Peyer's patches. Among kids slaughtered at 360 dpi, only one of those vaccinated prior to infection had lesions and these were only few focal granulomas whereas more severe lesions, and with no differences between groups, were found in those animals only infected or those vaccinated after the infection. According to these results, vaccination does not prevent infection but has a clear protective effect if administered before challenge; however, in most of the animals with an established infection, vaccination does not stop the progression of the lesions.

Keywords :
Vaccination, host response, pathology, experimental

O-01.11
PHENOTYPIC DIVERSITY IN THE IMMUNE RESPONSE AGAINST MYCOBACTERIUM AVIUM SUBSP. PARATUBERCULOSIS IN MAP-INFECTED DAIRY COWS

Meredith Frie[1], Kelly Sporer[2], Paul Coussens[2]

[1]Cell and Molecular Biology Program, Michigan State University, East Lansing, MI, USA.[2]Department of Animal Science, Michigan State University, East Lansing, MI, USA

Johne's disease is a serious illness affecting dairy herds in the United States. Johne's disease has a high economic burden: estimates suggest that Johne's disease costs the US dairy industry between $200 million and $1.5 billion every year. Mycobacterium avium subsp. paratuberculosis (MAP), the cause of Johne's disease, is widespread in the US. Recent estimates suggest that the US dairy herd prevalence of Johne's disease was approximately 90%. With no effective vaccines or treatments for Johne's disease, characterizing and understanding immune responses to MAP is critical for future control efforts. To capture a broad scope of immune responses to MAP antigens in vitro, peripheral blood mononuclear cells (PBMCs) from 203 healthy Holstein cows (Johne's negative) and 78 ELISA-positive Holstein cows (Johne's positive) were isolated. PBMCs were cultured in the presence of no stimulation, MAP antigenic stimulation or pokeweed mitogen (PWM) stimulation. Immune responses to MAP were characterized using two four-color flow cytometry stains covering the activation and phenotype of CD4+ helper T cells, CD8+ cytotoxic T cells, γδTCR+ T cells and surface IgM (SIgM)+ B cells. In both Johne's negative and Johne's positive groups, no B or T cell subsets showed significant activation in response to MAP stimulation when considered in total. However, a deeper analysis revealed a large degree of diversity in T cell responsiveness to MAP in culture: some cows demonstrated activation (>5% increase in activated cells), some demonstrated no response (less than a 5% increase or decrease in activated cells) and some demonstrated reduced activation (>5% decrease in activated cells). These trends were evident in Johne's negative cows and in Johne's positive cows. However, the proportion of cows with a positive T cell response was significantly higher in Johne's positive cows and the proportion of non-responding cows was significantly lower. These data suggest that the test-negative Johne's cohort includes either infected cows with currently undetectable antibody titers or cows that were transiently infected with MAP. In addition, the diversity of T cell responsiveness in Johne's positive cows strongly warrants further investigation to better understand the underlying causes of these diverse immune phenotypes in response to MAP.

Keywords : Adaptive immunity, T cells, host immunity

O-01.12
PARATUBERCULOSIS VACCINATED CATTLE LIFESPAN AND UNSPECIFIC PROTECTION

Juste RA[1,2], Vazquez P[1,3], Geijo MV[1], Serrano M[1], Elguezabal N[1], Molina E[1], Sevilla I[1], Alonso Hearn M[1], Perez V[4], Garrido JM[1]

[1]NEIKER-Tecnalia, Basque Institute for Agricultural Research and Development, Derio, Bizkaia, Spain. [2]SERIDA, Regional Service of Agricultural Research and Development, Villaviciosa, Asturias, Spain. [3]SALUVET, Department of Animal Health, Faculty of Veterinary Sciences. Complutense University of Madrid (CUM). [4]Department of Animal Health. Veterinary School. University of Leon. Leon, Spain

Mycobacterial vaccination is a controversial subject. In general, the emphasis for fighting TB was on diagnosis based on the grounds that identifying infected individuals and dealing with them, either by treatment in humans or by culling in animals was the most specific and efficient strategy. This perspective has also marked the approach to the other relevant cattle micobacteriosis, paratuberculosis, in spite of overwhelming evidence on vaccine efficacy. Since there is strong evidence of an unspecific effect on children culling by BCG vaccination, we wanted to know whether or not such an effect was also present in paratuberculosis vaccinated cattle.
Age at culling for 3547 cows vaccinated at over 3 months of age and 1731 of matched age but not vaccinated, and of 80 and 782 cows vaccinated within their 3 first months of life during a period of 9 years in 26 Friesian herds was recorded and submitted to survival analysis in one year intervals using the LIFETEST procedure of the SAS statistical package in order to compare cattle lifespan between groups. In another study on 987 Friesian slaughtered cows, three epidemio-pathogenic forms of paratuberculosis were defined. Age of these animals according to patency or not was submitted to a similar analysis in order to compare the effects of paratuberculosis on cattle lifespan and to estimate the frequency of clinical paratuberculosis incidence per year of age.
The Sidak logrank test showed no significant difference for cattle vaccinated after 3 months of age (p=0.3146), but highly significant for animals vaccinated earlier (p=0.0004). Maximum differences between vaccinated an unvaccinated animals occurred in animals between 1 and 4 years old, with 6.7%, 12.4%, 11.0% and 8.9% differences at each one year interval. Comparing with patent paratuberculosis, there was 6.7%, 9.8% and 1.8% less culling associated to vaccination at 1, 2 and 3 years of age than that associated to patent forms in the slaughterhouse study. These frequencies were considered as unspecific protection and represented a reduction of unvaccinated controls culling at those years of 31%, 19% and 2%. Paratuberculosis specific protection was 5% at 2 years of age and then 12%, 10%, 5%, 1% and 1% in the successive years.
These results confirm an association of paratuberculosis vaccination with extended cattle lifespan and an unspecific effect during the first years of life that is similar to that observed in human tuberculosis vaccination suggesting that collateral benefits of paratuberculosis vaccination might be a non-negligible part of its effects.

Keywords : Paratuberculosis, cattle lifespan, vaccination, unspecific protection

O-01.13
EARLY MARKERS FOR INFECTION AND VACCINATION AGAINST JOHNE'S DISEASE

Adel M. Talaat, Aubrey Berry[1]

[1]*University of Wisconsin-Madison, WI, USA*

Johne's disease, caused by Mycobacterium avium subspecies paratuberculosis (MAP) is a chronic gastroenteritis of ruminants. Although infection often occurs within the first few months of life, clinical signs do not appear until 2-5 years of age. Current diagnostic tests, such as fecal culture and ELISA, have poor sensitivity for detection of the early phase of Johne's disease (JD) and are not compliant with the principal of DIVA (differentiate infection from vaccinated animals). To better understand the early stage of JD and to identify a suitable approach for early detection, we employed RNASeq technology to profile the transcriptome of goats following infection with field isolate of MAP or following vaccination with a live-attenuated vaccine that we developed in our group. Peripheral blood mononuclear cells (PBMC's) were collected throughout one of our Johne's vaccine experiments in goats. The PBMC transcriptomes of goats were profiled to evaluate differential gene expression between a subset of samples from either 30 days post-vaccination, 30 days post-infection, or a naive, non-infected control group. Preliminary results on differential gene expression indicate 88 significantly differentially expressed genes out of 11,009 genes between goats at 30 days post-infection and the naïve, non-infected controls. Interestingly, the 30 days post-vaccination group had more differentially expressed genes (720 out of 10,985 and 746 out of 11,099) compared to the non-infected control and the 30 days post-infection groups, respectively. Preliminary evaluation of the significantly differentially expressed genes indicated a large number of genes with immunological and inflammatory functions, including IL-18 binding protein, IFN-γ, IL-17A, and IL-22. Identifying genes involved in the host response to disease is critical to understanding the pathophysiology of MAP infection and identification of specific biomarkers for early infection or vaccination. Currently, we are working on identifying similar markers in cows to be used for understanding JD pathogenesis and developing a DIVA-based assay.

Keywords :
Paratuberculosis, Transcriptome, Early Infection, Early Detection, DIVA

O-01.14
EVIDENCE AND DETECTION OF SHEDDING PATTERNS OF MYCOBACTERIUM AVIUM SUSBSPECIES PARATUBERCULOSIS IN DAIRY CATTLE

R. Guatteo[1], P. Blanquefort[2], A.Delafosse[2], A. Joly[2], E. Meens[2], C. Fourichon[1]

[1]*Oniris, UMR Oniris-INRA 1300 BioEpAR, Nantes, France.* [2]*GDS Grand Ouest, France*

In herds infected by Mycobacterium avium subspecies paratuberculosis (Map), the early detection and culling of animals likely to be persistent heavy shedders could be a key factor for limiting Map transmission. The literature provides scarce data on shedding over a long period. The objective was to describe the heterogeneity of Map shedding over time in cattle to characterize different shedding patterns of interest.

A one year longitudinal follow-up was performed in 22 French dairy herds located in Normandy known to be infected (> 8% of seropostivie cows on the previous year). Within each herd, cattle (>12 months) were sampled four times every 4 months. At each sampling time, feces and blood samples were collected to be tested using respectively qPCR and blood ELISA. A similar approach was conducted on a subsample of 200 cows (selected on the basis of previous PCR results) to describe the shedding pattern on a 7-days period (3 samples within a week). The results were described for each sampling time and over time to identify shedding pattern, based on both qualitative (negative vs positive) and quantitative approaches among positive samples (Ct < 35 or 35ffiCt<40). In a second step, we investigated the possibility to discriminate these longitudinal shedding patterns on a single sampling time based on the combination of both qPCR and serology.

Finally, 1137 Hosltein and Normande cows were sampled. Among them, 6% were considered as persistent heavy shedder while 18% were considered as frequent heavy shedder. Only 13% of the studied cows never shed during the study period. Similar shedding patterns were observed within a 7 days period. To discriminate heavy persistent shedder from intermittent shedder, a combination of a single positive qPCR result (Ct<35) associated with a concomitant ELISA positive result (S/P ratio > 90) was highly specific and allowed to identify around 50% of heavy persistent shedding cows.

To our knowledge, this is the first study describing at such a scale the different Map shedding patterns. These results could contribute to the prioritization for culling in dairy herds or the identification of phenotypes of interest for further genomic studies.

This study was funded by INRA (métaprogramme GISA), Apis-Gene, and GDS France.

Keywords :
Dairy cattle, Shedding patterns, Phenotypes

O-01.15

Composite Immune Biomarkers Associated with Resistance or Susceptibility to Johne's Disease: Data from a Deer Model

Frank T Griffin[1], Colin G Mackintosh[2], Rory O'Brien[1], Simon Liggett[1], Liam Brennan[1]

[1]*Disease Research Laboratory, University of Otago, Dunedin, New Zealand.* [2]*Invermay Agricultural Centre, Mosgiel, New Zealand*

While it is widely accepted that the host genotype plays an important role in Resistance (R) or Susceptibility (S) to infectious diseases, the genes that define these traits have not been properly categorised. As the response is multigenic and likely to be linked with immune competence, we have explored a systems approach that targets key immunological pathways, rather than individual genes, or gene wide associations.

During the past 10 years our laboratory has carried out extensive studies on pure bred lines of deer, shown to be naturally resistant or susceptible to MAP infection under field conditions. Progeny (100+) from individual sires have been experimentally infected with virulent MAP. Their phenotypes were confirmed retrospectively following necropsy. Animals were slaughtered at the onset of clinical disease (Susceptible - S) or electively (Resistant - R) around 9 months post infection. A panel of 5 innate biomarkers (IL-1A, IL-12A, IL-23p19, NOS2 & SOCS3) were upregulated in blood monocyte derived macrophage cultures, infected by MAP and derived from S animals. Though informative, this method was cumbersome, as it required large blood volumes (>200mL) and was impractical as a diagnostic method to characterize the disease phenotype of individual animals.

An alternative peripheral blood mononuclear culture system was developed that required much smaller (20mL) blood samples to monitor expression of innate or adaptive immune biomarkers. Cells from R and S animals, stimulated either with MAP or a Polyclonal activator (Staphylococcal Enterotoxin B – SEB), expressed different levels of a separate panel involving 5 innate and adaptive biomarkers, which may be used to distinguish between R and s animals. Gene expression levels were similar in samples obtained prior to experimental infection or up to 9 months post infection. Segregation of marker expression was also seen at a genotypic level in uninfected animals with predicted R or S phenotypes, and the response was moderately heritable (0.3 +/- 0.06). The overexpressed genes identified in R or S progeny initially, and now being tested as predictive markers in 40 young crossbred deer, representative of sires and dams used widely for commercial deer farming in New Zealand.

Keywords :
Immune Biomarkers, Selection for, Susceptibility and Resilience , Johne's Disease, Deer

O-01.16

Whole genome association analysis of resistance / susceptibility to paratuberculosis in French Holstein and Normande cattle

Sanchez MP[1], Guatteo R[2], Davergne A[3], Grohs C[1], Capitan A[1,4], Blanquefort P[5], Delafosse A[6], Joly A[7], Ngwa-Mbot D[8], Biet F[9], Fourichon C[2], Boichard D[1]

[1]*GABI, INRA, AgroParisTech, Jouy en Josas, France.* [2]*UMR Oniris-INRA 1300 BIoEpAR, Nantes, France.* [3]*Groupement de Défense Sanitaire Haute-Normandie, Rouen, France.* [4]*Allice, Paris, France.* [5]*Groupement de Défense Sanitaire Pays de la Loire, Trelazé, France.* [6]*Groupement de Défense Sanitaire Orne, Alençon, France.* [7]*Groupement de Défense Sanitaire Bretagne, Vannes, France.* [8]*GDS France, Paris, France.* [9]*ISP, INRA, Nouzilly, France*

The purpose of this study was to identify genomic regions associated with resistance / susceptibility to Mycobacterium avium ssp. paratuberculosis (MAP) infection in French Holstein and Normande cattle. A case-control genome-wide association study (GWAS) was performed. Cases were infected shedder cows (confirmed with both positive blood ELISA and positive PCR on feces) or clinical cases. The control population included only animals with three repeated negative blood ELISA, negative fecal PCR test and at least 72 months old. To limit exposure biases, the control animals were required to be born in the same herd and at the same time period as confirmed MAP positive cows (shedder or clinical). A total of 405 Normande (210 cases and 195 controls) and 989 Holstein (437 cases and 552 controls) cows were thus genotyped with the Illumina BovineSNP50 BeadChip (39,055 informative markers). GWAS was conducted within breed with GCTA software, accounting for the population structure through a 50K-based genomic relationship matrix.

The most significant region associated with infectious status was found on chromosome (BTA) 12 at 69.8 Mb (pvalue=2E-8) in Holstein cows. Another association was identified on BTA23 in the region of the major histocompatibility complex (MHC) in both Normande (27.8 Mb; pvalue=4E-5) and Holstein cows (25.6 Mb; pvalue=4E-6). Other chromosomal regions (pvalue < 7E-5) were found on BTA1 (55 Mb), BTA10 (91 Mb) and BTA13 (56.6 Mb) in Holstein and on BTA1 (92.6 Mb), BTA9 (39.3 Mb), BTA15 (58.4 Mb) and BTA17 (8 Mb) in Normande cows. Several additional regions were found with more moderate significance level (pvalue < 1E-4) in Normande (BTA13 and 25) and Holstein (BTA3, 9, 11, 25, 26 and 28) cows.

This on-going study presents encouraging results. Additional cases and controls will be collected and genotyped in 2016. In order to directly pinpoint candidate causal mutations, whole genome sequences of the cows will then be imputed using the 1000 bull genome reference population and GWAS will be carried out directly on whole genome sequence data.

This study was funded by INRA (métaprogramme GISA), Apisgene, and GDS France. The authors thank the farmers and GDS who contributed to this project.

Keywords :
Paratuberculosis, genetic resistance, bovine

O-01.17

GENE EXPRESSION ANALYSIS IDENTIFIES GENES DIFFERENTIALLY REGULATED IN BOVINE AND OVINE MODELS OF MYCOBACTERIUM AVIUM SUBSPECIES PARATUBERCULOSIS EXPOSURE

Auriol Purdie[1], Kumudika de Silva[1], Douglas Begg[1], Karren Plain[1], Richard Whittington[1]

[1]*School of Life and Environmental Sciences, Faculty of Veterinary Science, The University of Sydney, Australia*

Microarray chip technology is a tool for probing the expression of thousands of genes in a single experimental process. It can provide an overview of gene expression patterns in an animal during an infection. The aims of this study were to individually evaluate gene expression (transcriptomic) changes 2 to 15 months post MAP exposure in at least 20 each of Merino sheep and other breeds (Border Leicester, Poll Dorset and Suffolk x Merino) and Holstein / Holstein Red cattle exposed to an appropriate dose of MAP (C or S strain). Gudair® vaccinated sheep were also studied. These were compared to age-matched controls over multiple timepoints close to MAP exposure, and with reference to clinical outcome at a later time, as defined by gross pathology, histopathological lesions, faecal cultures and tissue culture.

Bovine Affymetrix GeneChip microarray and extensive ontological analysis identified differentially regulated groups of genes associated with disease outcomes. These gene lists were mined to determine biological context. We found associations with both IFNg and MHC pathways and a range of previously unknown pathways. The results suggest that even at early timepoints post exposure to MAP there are mechanisms occurring that may potentially be manipulated to alter the eventual clinical outcome or that may be monitored to predict the course of the disease within the exposed animals.

Keywords :
Gene expression, pathogenomics, bioinformatics, disease pathogenesis

O-01.18

OVINE MACROPHAGE-MEDIATED KILLING OF MYCOBACTERIUM AVIUM SUBSPECIES PARATUBERCULOSIS IN VITRO IS ENHANCED BY VACCINATION

Hannah B Pooley[1], Kumudika de Silva[1], Auriol C. Purdie[1], Douglas J. Begg[1], Richard J. Whittington[1], Karren M. Plain[1]

[1]*School of Life and Environmental Sciences, Faculty of Veterinary Science, The University of Sydney, Camden, NSW, Australia*

This work examines the hypothesis that final clinical outcomes of lengthy ruminant Mycobacterium avium subspecies paratuberculosis (MAP) infection trials could be correlated with the in vitro ability of peripheral blood mononuclear cells (PBMCs) to kill MAP after exposure. If this is indeed true, in vitro infection assays could become a rapid and cost effective alternative to animal trials for the screening of vaccine candidates. This assay examines adherent monocytes that are infected with MAP in vitro and are then co-cultured with autologous lymphocytes (operationally these are non-adherent PBMCs). The viability of intracellular MAP is then determined using a novel method developed based on the relationship between the rate of MAP growth and the concentration of live MAP at the start of culture. The assay was optimised for incubation time, multiplicity of infection and culture duration and then a pilot study was carried out using cells from Gudair® vaccinated MAP exposed diseased (histological lesions with a score of Perez type 3) and non-diseased (no histopathological lesions) sheep as well as vaccinated and non-vaccinated unexposed control sheep. The MAP exposed sheep had been orally challenged with MAP 12 months prior to the collection of cells for the in vitro assay. The infection outcomes were determined at 13 months after oral exposure. Two sheep were used for each vaccinated treatment group and four sheep for the non-vaccinated controls. The assay was carried out in duplicate for each animal. Results of the in vitro assay showed a large variation in the killing ability of PBMCs from non-vaccinated control animals. In contrast, vaccination decreased variability between animals and increased in vitro killing of MAP by macrophages. Although the killing ability of PBMCs from vaccinated sheep was enhanced, we were unable to differentiate between diseased and non-diseased individuals, so further testing of a larger number of animals is planned. However, the assay developed has proven to be a useful tool for examining the host response to vaccination. Continual use of the assay to test earlier infection time points would further develop our understanding of the effects of vaccination on the ability of PBMCs to kill MAP.

Keywords :
In vitro, assay, viability, vaccination, infection, Mycobacterium avium subspecies paratuberculosis, Peripheral blood mononuclear cells

POSTER

P-01.1

Discovery of sero-reactive antigens in Mycobacterium avium subsp. paratuberculosis with protein microarrays

Lingling Li[1], John P Bannantine[2], Craig A Praul[1], Juan Antonio Raygoza Garay[1], Murray E HInes II[3], Robab Katani[1], Michael Mwangi[1], **Vivek Kapur**[1]

[1]Pennsylvania State University, State College, Pennsylvania, USA. [2]National Animal Disease Center, USDA-ARS, Ames, Iowa, USA. [3]Veterinary Diagnostic and Investigational Laboratory, The University of Georgia, Tifton, Georgia, USA

Mycobacterium avium subspecies paratuberculosis (MAP) is the causative agent of Johne's disease (JD), a chronic intestinal disease of domestic animals. Methods for the early detection of infected animals are currently lacking, and limit the ability to control spread of the disease. To identify possible biomarkers for early (even prior to fecal shedding) MAP infection, we developed a protein microarray with individual recombinant MAP proteins as a tool for sero-diagnostic antigen discovery by studying the dynamics of the humoral immune response in a baby goat model of infection from JDIP/MDA vaccine project. A protein microarray consisting of ~600 recombinant MAP proteins was constructed and used to identify reactive antigens in serum samples from experimentally MAP-infected (n=10) and negative control (n=10) goats at 7 time-points (0 to 12 months post infection) that were part of the JDIP coordinated vaccine trial program (Front Cell Infect Microbiol, 4:26, 2014). Our preliminary analysis has resulted in the identification of ~50 MAP recombinant proteins that are sero-reactive in infected animals, but not in the controls. Eight of these were detected as early as 2 months post challenge, which is considerably sooner than the detection of fecal MAP shedding (average 6.9 months post infection) by fecal culture and/or PCR in these same animals. Further, our preliminary analyses suggest that individual MAP recombinant proteins were reactive even in MAP-infected goats that remained serologically negative on commercially available MAP ELISA testing, suggesting that the protein microarray approach may enable the identification of antigens that are sero-reactive even during the early (pre-shedding) stages of infection. Future studies with larger numbers of recombinant proteins and serum samples are planned and it is expectable that identification of validated MAP antigens enables the development of the next generation of sensitive and specific diagnostic assays for detection of early infection.

Keywords :
Mycobacterium avium subsp. paratuberculosis (MAP), protein microarray, sero-reactive antigens

P-01.2

Evaluation of the cellular immune response in naturally infected cattle with Mycobacterium avium spp. paratuberculosis from Argentina

Moyano Roberto Damián[1], Romero Magalí[2], Alvarado Pinedo Fiorella[2], Gravisaco María José[1], **Santangelo María de la Paz**[1], Travería Gabriel E[2], Romano María Isabel[1]

[1]Instituto de Biotecnologia, CICVyA INTA, Buenos Aires, Argentina. [2]CEDIVE y Facultad de Ciencias Veterinarias, UNLP, Argentina

Introduction: Mycobacterial infections represent major health problems, both in humans and farm animals. Mycobacterium avium subsp. paratuberculosis (MAP) is the causative agent of Johne_disease in ruminants such as cows and sheep causing chronic intestinal inflammation. Exposed animals enter a subclinical period of 2 to 5 years after which a proportion of the infected animals develop severe enteritis with chronic diarrhea.Objectives. Study the T lymphocytes response in naturally infected cattle with MAP in order to detect infection or protection markers.Materials and Methods. 9 female calves, born in a naturally infected dairy herd with MAP, were sampled over a period of 18 months. 9 female calves were sampled from a free TB and PTB herd as control. Blood samples and fecal material were collected every two months. The presence of MAP in positive fecal cultures was confirmed by amplification of the MAP specific IS900 by qPCR. Peripheral blood T lymphocytes (CD4+CD25+, CD8+CD25+, WC1+CD25+) stimulated for 24h with a protein extract of Mycobacterium avium (PPD-a, BOVIGAM®), were evaluated by flow cytometry.Results. The presence of MAP in fecal cultures was confirmed in 5/9 animals from the positive herd by qPCR with primers IS900 at 14 months of age. This result was accompanied by a high response of gamma-delta and CD8+ cytotoxic lymphocytes. The cytokines profile was evaluated by RT-qPCR, resulting in an increase of IFNg in 2 of the animals at 16 months of age. These preliminary studies could contribute to the development of new diagnostic techniques and the identification of infection or protection markers.

Keywords :
Immune response, Infection, Protection Markers

P-01.3

FIBROGENIC AND INFLAMMATORY RESPONSE TO MYCOBACTERIUM AVIUM SUBSP. PARATUBERCULOSIS IN BOVINE ILEAL SUB-EPITHELIAL FIBROBLASTIC POPULATIONS

Elisabetta Stefani[1], Annalucia Tondo[1], Michele Gastaldelli[1], Andrea Matucci[1], Giulietta Minozzi[2], Venera Russo[3], Ignazio Castagliuolo[3], **Nicola Pozzato**[1]

[1]Istituto Zooprofilattico Sperimentale delle Venezie, Sezione di Verona, Verona, Italy. [2]Parco Tecnologico Padano, Lodi, Italy. [3]Dept. of Molecular Medicine, University of Padova, Padova, Italy

Latent Mycobacterium avium subsp. paratuberculosis (MAP) infection evolves toward multifocal or diffuse granulomatous lesions in which intestinal fibrosis, caused by activated fibroblasts, become apparent. Nonetheless, it is still unclear whether fibroblasts activation relies on local inflammation or on the direct interaction with MAP. In this study we isolated 4 primary fibroblastic cell populations from the ileum of paratuberculosis uninfected cows. Cell cultures resulted to be Desmin negative and Vimentin positive and were composed by fibroblasts α-SMA(-) and myofibroblasts α-SMA(+) population (FMP). α-SMA(+) cells showed the capability to act as non-professional antigen presenting cells (APC) as following IFN-γ stimulation significantly increased the expression of MHC-I, MHC-II, CD80 and CD86 molecules, by cytofluorimetric analysis.To asses the inflammatory response in FMP we exposed our cell cultures to mycobacteria (Mycobacterium avium avium and 2 different MAP field strains: MAP993 and MAP445) and LPS.
We observed an increase of IL-6 expression in FMP exposed to LPS and MAP 993, while IL-8 expression increases only with LPS at 48 h. A significant increase in the expression resulted for MMP1 gene with both MAP strains, while LPS induced a marked increase for MMP1, MMP2 and MMP14.To deeply investigate MAP-FMP and LPS-FMP interactions we analyzed the whole transcriptome of the 4 lines by mRNA-sequencing after stimulation with MAP 993 and LPS (at 4, 24 and 48 hours). Smear plot analysis showed that LPS induces a marked increase in the differentially expressed (DE) genes in all cell lines. On the other hand MAP 993 strain induce an increase in the expression of the DE genes in a limited number of genes, and a differential expression among cell lines. Among these only one gene, IL-6, codes for a cytokine involved in the inflammatory response and none is involved in the fibrogenic response.In conclusion, fibroblastic population are able to directly interact with MAP that elicits a mild inflammatory response and no fibrogenic response. On the contrary LPS is a potent activator of these populations in both inflammatory and fibrogenic responses.

Keywords :
Host response, fibroblast, inflammation, fibrogenesis

P-01.4

DIVERGENT CELLULAR RESPONSES DURING ASYMPTOMATIC SUBCLINICAL AND CLINICAL STATES OF DISEASE IN COWS NATURALLY INFECTED WITH MYCOBACTERIUM AVIUM SUBSP. PARATUBERCULOSIS

Judy Stabel[1], J. P. Bannantine[1]

[1]USDA-ARS-National Animal Disease Center, USA

Infection of the host with Mycobacterium avium subsp. paratuberculosis (MAP) results in a chronic and progressive enteritis that traverses both subclinical and clinical stages. The mechanism(s) for the shift from asymptomatic subclinical disease state to advanced clinical disease are not fully understood but host immunity is decidedly a factor. Th1-mediated immunity dominates the early subclinical stage of infection with Th2 dominated immunity observed during clinical disease. However, this paradigm is an oversimplification as significant overlap of Th1/Th2 immunity exist in infected animals, even those that are eventually culled because of advanced clinical disease. More recently, CD4+ Th17-mediated immunity has been proposed as a significant immune mediator in mycobacterial infections. In the present study, naturally infected dairy cattle were defined as either subclinical and clinical infection groups, along with noninfected control cows of similar parity to study host immune responses in different stages of infection. Percentages of CD4+ and CD8+ T cells within freshly isolated PBMCs were lower in cows with clinical disease compared to subclinical cows. Interestingly, although the percentage of $\gamma\delta$TCR+ T cells was also reduced in total PBMCs from clinical cows, antigen stimulation of cells provoked an increase in activation markers on this cell subset. Both infection groups had higher secretion of IFN-γ, TNF-α, and IL-2, whereas only clinical cows had increased secretion of IL-10, IL-12 and IL-18 upon stimulation of cells with antigen. Conversely, secretion of IL-17α was decreased for clinical cows. The majority of differences in cytokine gene expression between subclinical and clinical disease states were increased IFN-γ, IL-12, IL-18, RANTES, IL-23, and iNOS, as well as IL-4 noted for subclinical cows. Increased IL-10 and IL-17 gene expression were observed for both infection groups compared to the control noninfected cows. We reason that a complex coordination of immune responses occurs during MAP infection, with these responses shifting as the host transitions through the different stages of infection and disease (subclinical to clinical). Further understanding of this series of events as characterized by Th1/Th2/Th17 responses will provide knowledge of Johne's Disease progression and may direct insightful intervention strategies.

Keywords :
Cattle, Immunity, Cytokine

P-01.5

Cellular responses to Mycobacterium avium, subsp. paratuberculosis in colostrum-deprived and colostrum-replete Holstein calves supplemented with fat-soluble vitamins

Judy Stabel[1], L. A. Krueger[2], C. Beitz[3], B. Humphrey[1]

[1]*USDA-ARS-National Animal Disease Center, Ames, IA.* [2]*Iowa State University, Ames, IAD.* [3]*Iowa State University, Ames, IAS, USA*

Immune benefits of colostrum are attributed to passively transferred IgG but also to growth factors, cytokines, antimicrobial peptides, and leukocytes. Non-nutritive compounds in colostrum promote Th2-biased immune responses to early microbial encounters and prevent harmful, inappropriate inflammation in neonatal tissues. Post-natal, nutritional influences such as vitamin A, D3, or E deficiencies may cause dysregulation of immune signaling and decreased integrity of immune responses, thus compromising neonatal health. The present study was conducted to determine if neonatal bovine leukocyte subsets recognize and respond to MAP antigens within the first 14 d of age when immunity of the calf is still partially naive. The study was designed to test the hypothesis that colostrum, vitamin supplementation, or both nutritional treatments alter mononuclear cell function and subsequent responses to MAP antigens. Peripheral blood mononuclear and mesenteric lymph node cells (PBMC and MNL) were obtained from 30 calves that were assigned randomly at birth to treatments: 1) colostrum deprived (CD), 2) colostrum replacer (CR), 3) CR, vitamin A; 4) CR, vitamin D3; 5) CR, vitamin E; 6) CR, vitamins A, D3, E, in a 14 d study. Calves were injected with vitamin supplements and fed pasteurized whole milk (PWM; CD calves) or fractionated CR at birth. Thereafter, all calves were fed PWM fortified with vitamins according to treatment. Calves were orally inoculated with 108 cfu of Mycobacterium avium subsp. paratuberculosis (MAP) on d 1 and 3. PBMC and MNL were analyzed by flow cytometry as fresh cells, after 3 d culture with PHA, and after 6 d culture with a whole cell sonicate of MAP. Peripheral γδ T cells were identified as a predominant lymphocyte subset, with decreased percentages noted in CD calves. Stimulation of PBMC with PHA increased CD4 and CD8 subsets, whereas the MNL response was dominated by expansion of B cells. PHA and MPS stimulation decreased the relative abundance of γδT cells among PBMCs, but MNL γδT cells increased upon stimulation with MPS. These results identify γδT cells as key early responders to intracellular infection in neonatal calves and suggest that colostrum may be an important mediator of this response.

Keywords :
Colostrum, Calves, Vitamins

P-01.6

Association of immunological gene expression profiles of intestinal tissues to Mycobacterium avium subspecies paratuberculosis infection levels of calves

Christina Chemnitz Grønbæk[1], Heidi Mikkelsen[1], Louise Brogaard[1], Aneesh Thakur[1], **Gregers Jungersen**[1].

[1]*Technical University of Denmark, National Veterinary Institute, Section for Immunology and Vaccinology, Copenhagen, Denmark*

The aim of the study was to describe immunologically relevant gene expression patterns in intestinal tissues related to different levels of Mycobacterium avium subspecies paratuberculosis (Map) infection. Our major focus area was to quantify the magnitude of Th17 related immune responses during Map infection compared to Th1 responses. For this study we retrospectively selected samples of 6 calves that were all experimentally infected with Map at two weeks of age and based on serology, histology and Map tissue load were classified as protected (n=2) or unprotected (n=2) after vaccination, or un-vaccinated infected controls (n=2). From each calf, 7 intestinal tissue samples and 3 lymph node samples, collected at 10 months of age, were used for cDNA synthesis. Parallel expression analysis of a total of 37 selected genes including inflammatory, Th1 and Th17 related genes was performed by high-throughput reverse transcriptase (RT) qPCR on the Fluidigm (GenEx) platform.The results showed that Map infection, as expected, leads to increased expression of local IFN-γ. Expression of IL-10 also increased as a result of Map infection, and this increase was more correlated to the Map tissue load than IFN-γ, indicating a shift towards a regulatory environment as infection progress. Th17-mediated immune responses were suppressed at this stage. Gene expression of all other genes could not be interpreted in relation to infection status.High throughput RT qPCR can be used for exploring gene expression patterns in response to Map infection but larger study groups are needed to fully understand which are key mechanisms and pathways responsible for protection or disease.

Keywords :
Gene expression, immune response, Th17 responses, Map infection, protection

P-01.7

FREE-LIVING AMOEBAE AS A PHAGOCYTOSIS MODEL OF MYCOBACTERIUM AVIUM SSP. PARATUBERCULOSIS

Etienne Robino[1], **Ascel Samba**[1], Thierry Cochard[2], Maxime Branger[2], Franck Biet[2], Yann Héchard[1]

[1]UMR 7267 CNRS Ecologie et Biologie des Interactions (EBI), Université de Poitiers, 86073 Poitiers. France. [2]INRA, UMR1282, Infectiologie et Santé Publique, INRA centre Val de Loire, F-37380 Nouzilly, France.

Background: Free-living amoebae (FLA) are protozoa found in water and soil that graze on bacteria and digest them by phagocytosis. In some cases, bacteria are able to resist phagocytosis and eventually multiply within FLA (e.g. Legionella pneumophila). Therefore FLA are described as potential reservoirs of pathogenic bacteria. It has been shown that some Mycobacterium were able to resist FLA phagocytosis. Among mycobacteria, Mycobacterium avium subsp. paratuberculosis (Map) takes a special place in veterinary medicine since this mycobacteria is the causal agent of paratuberculosis in ruminants. The infectious processes of Map and the transmission to the ruminant host are complex and remain mainly unknown. During the infectious cycle Map is excreted and may persists in the environment. The Map interactions with the soil components, including amoebae have been describing in few studies and suggested that Map may resist and persist in these organisms. Objective: to characterize at the cellular level the interaction of Map with free-living amoebaeMethod: Infection of Acanthamoeba will allow to follow the fate of Map after phagocytosis. Also the comparison between several strains of Map including field isolates with different host origins and genotypes will help to see if they share the same behaviour. Finally, the expression of selected genes, likely involved in phagocytosis modulation, will be assessed along the infection process. The final goal is to assess the role of FLA in Map survival and transfer in the environment.

Keywords :
Amoebae, interactions, phagocytosis

P-01.8

VARIATIONS IN T CELL TRANSCRIPTION FACTOR GENE STRUCTURE AND EXPRESSION LEVELS ASSOCIATED WITH THE TWO DISEASE FORMS OF OVINE PARATUBERCULOSIS

Nicol Louise[1], Wilkie Hazel[1], Gossner Anton[1], **Watkins Craig**[2], Dalziel Robert[1], Hopkins John[1]

[1]The Roslin Institute & R(D)SVS, University of Edinburgh, Easter Bush, Midlothian, EH25 9RG. UK. [2]Moredun Research Institute, International Research Centre, Pentlands Science Park, Penicuik, Midlothian EH26 0PZ, UK.

Two different forms of clinical paratuberculosis in sheep are recognised, related to the level of bacterial colonization. Paucibacillary lesions contain lymphocytes and few bacteria, and multibacillary pathology is characterized by infiltration with heavily-infected macrophages1. Analysis of cytokine transcripts has identified inflammatory Th1/Th17 T cells as critical to the development of paucibacillary disease and Th2 cytokines to be correlated with the multibacillary form2.
The master regulator transcription factors TBX21, GATA3, RORC2 and RORA are critical for the development of these T cell subsets3. Sequence variations of the transcription factors have also been implicated in the distinct disease forms of human mycobacterial and gastrointestinal inflammatory diseases4-5. Relative RT-qPCR was used to compare expression levels of each transcript variant of the master regulators in the ileocecal lymph nodes of uninfected controls and sheep with defined paucibacillary and multibacillary pathology. Low levels of GATA3 in multibacillary sheep failed to confirm that multibacillary paratuberculosis is caused by a Th2 immune response. However, high levels of TBX21, RORC2 and RORC2v1 highlights the role of Th1 and Th17 activation in paucibacillary disease. Increased RORAv1 levels in paucibacillary tissue emphasises the role of RORα in Th17 development; while elevated levels of RORAv4 in multibacillary sheep hints that this variant might inhibit RORα and depress Th17 development.References:Clarke CJ.;J Comp Pathol 1997;116(3):217-61.Robinson MW, et al.;Infect Immun 2011; 79(5):2089-97.Zhu J, et al.; Annu Rev Immunol 2010; 28:445-89.Kanai T, et al.; Mucosal Immunol 2012;5(3):240-7.Rauen T, et al.; Genes Immun 2012;13(4):346-50.Acknowledgements: We thank Dr Francesca Chianini, The Moredun Research Institute (MRI) for pathological diagnosis of diseased animals, MRI Bioservices and Joan Docherty (The Marshall Building, University of Edinburgh) for animal husbandry of infected and control animals and Joyce McLuckie, MRI, for bacterial culture.

Keywords :
T cell transcription factors, ovine, paratuberculosis

P-01.9
Early Interactions between Bovine Monocyte Derived Macrophages and Mycobacterium avium subsp. paratuberculosis

Heather Mathie[1], Joanne Stevens[1], Elizabeth Glass[1], Jayne Hope[1]

[1]*The Roslin Institute, The University of Edinburgh, Edinburgh, Scotland*

Mycobacterium avium subsp. paratuberculosis (MAP) infection of cattle has a damaging economic impact and is a cause for concern regarding potential zoonotic transmission. There are significant difficulties regarding preventing, detecting, and eradicating MAP infection in cattle herds, many of which stem from the ability of MAP to evade the host immune response (and thus diagnosis). This evasion is associated with the ability of MAP to persist within macrophages for prolonged periods of time. The precise mechanisms of this host-pathogen interaction are not fully understood; we hypothesise that the earliest interactions between MAP and host macrophages, including the mechanism of uptake, may play a pivotal role in determining the outcome of infection. In order to investigate this, we infected bovine monocyte derived macrophages (MDM) with two strains of MAP (the well-established K10 strain and strain C49, a recent clinical isolate from cattle) in order to model the infection in vitro. Following infection of MDM with the C49 strain, numbers remained fairly constant over a 24 hour time-course, whereas the K10 strain was completely eradicated by 24 hours post infection; this difference in survival is indicative of a loss of virulence in the K10 strain through lab adaptation. Due to the difference observed between the two strains in their ability to survive intracellularly, the early immune response of MDMs to each strain was analysed, shedding light on mechanisms which could be important for early clearance. By assessing cell surface molecule expression, gene expression, cytokine secretion, nitrite and reactive oxygen species production and phagosome acidification, a detailed picture of the early macrophage response was formed for both strains. We also demonstrated that the presence of serum antibody responses to MAP can impact uptake and survival of MAP within macrophages and impact their downstream effector functions. This suggests that the mechanism of uptake, for example via Fc or complement receptors, could play a role in determining downstream responses to MAP infection.

Keywords :
Macrophage, Innate response, MAP-host Interaction

P-01.10
Mucosal-Associated Invariant T (MAIT) cells in cattle with and without Johne's disease

Darcy Fletcher[1], Henao-Tamayo Marcela[1], MD PhD, Hess Ann M.[1], PhD, Garry Franklyn B.[1], DVM MS, Ordway Diane J.[1], PhD, Eckstein Torsten M.[1], MD PhD

[1] *Colorado State University, USA*

Johne's disease, caused by Mycobacterium avium subspecies paratuberculosis, is a chronic infectious disease of the intestine in ruminants divided into four stages: (1) silent stage, (2) subclinical stage, (3) clinical stage, (4) advanced clinical stage. The majority of studies have assessed the immune response in the last two stages of disease, but have not evaluated these responses during the beginning stages. It remains unknown how during early infection the host controls the infection in the micro-environment of the intestine.One of the potential protective mechanisms is the MR1-restricted MAIT system, a phenotype, which becomes depleted from the blood over time. It is not known when this process starts, when it ends or how it is initiated and regulated. Thus, it is important to understand the immune dysregulation, which occurs with MAP colonization and disease progression. Moreover, knowledge on the number, cell function and location of MAIT cells in relation to other cellular phenotypes may provide information required for vaccine development. MAIT cells play a regulatory role and are restricted to receiving their specific antigens from MHC related 1-bearing immune cells, primarily B cells. This places MAIT cells in the group of restricted T cells such as invariant NK cells and CD1-restricted T cells. In Mycobacterium tuberculosis infections, the proportion of MAIT cells in peripheral blood is inversely correlated to disease status of the host. MAIT cells were also found to be numerically and functionally deficient, and these deficiencies most likely contribute to immune system dysregulation at the site of infection.Our aim was to study this phenomenon to better understand what causes the decline of MR1-restricted MAIT cells. Most of the resources were developed for the human model and no cross-reactivity with other animal models were yet reported. Here we present our findings on the cross-reactivity with the bovine system as well as percentages of MAIT cells in local immune compartments of healthy and clinical cattle infected with Johne's disease including peripheral blood, small intestine, liver, spleen, and lymph nodes.

Keywords :
Immunology, Cattle, Host Response, MAIT cells

P-01.11

JOHNES DISEASE BOVINE MONOCYTE-DERIVED MACROPHAGES ARE INSENSITIVE TO EX VIVO INFECTION BY MYCOBACTERIUM AVIUM SSP. PARATUBERCULOSIS

O. Ariel[1,2], N. Gévry[2], E.M. Ibeagha-Awemu[1], **N. Bissonnette**[1]

[1]Agriculture and Agri-Food Canada, Dairy and Swine Research and Development Centre, 2000 College Street, Sherbrooke, Quebec, J1M 0C8 Canada. [2]Department of Biology, Université de Sherbrooke, Sherbrooke, Quebec, J1K 2R1 Canada

Bovine paratuberculosis (PTB), also known as Johne's disease, is a worldwide insidious disease of ruminants that is induced by an intracellular pathogen called Mycobacterium avium ssp. paratuberculosis (MAP). There is no cure for PTB, and vaccine therapy does not prevent new infection. The host's genetic susceptibility, tolerance to the intracellular pathogen during the early stage of the infection, and inefficient immune response are predisposing factors to the development of PTB. The pathogen invades the bovine intestinal tract, where it is phagocytosed by macrophages, the primary reservoir of MAP. It is therefore interesting to study this specialized immune cell type. Six PTB-positive cows identified by bacterial culture of fecal samples and six herdmates found to be negative by both serum ELISA and MAP culture were selected. Monocyte-derived macrophages from these cows were infected ex vivo for 4 and 24 h with live MAP. RNA was harvested, and the whole transcriptome was sequenced with 100-bp-end sequencing using Illumina HiSeq 2000 technology. A minimum of a 2-fold change in gene expression and a P-value ﬃ 0.05 after false discovery rate correction were used to select significant genes. In total, 243 and 364 genes were differentially expressed in primary bovine macrophages between the negative and positive cows at the 4- or 24-h post-infection time points, respectively. Common to both time points, the 37 upregulated and 33 downregulated genes were associated with the immune system, biological regulation, developmental processes, or the response to stimulus. The impact of ex vivo infection by MAP on the macrophages was observed mainly in the negative cows. In total, 712 and 2662 genes were identified as being differentially expressed between the control and the 4- and 24-h post-infection time points, respectively. Interestingly, no significant difference was observed for the positive cows. These findings suggest that monocyte-derived macrophages could be metabolically predisposed to not respond to MAP infection. These results give new insight into the possible effect of the predisposition of PTB cows on the ability of macrophages to achieve an efficient immune response to MAP infection, on putative outcomes in tissues, and on disease progression. Further research is required to determine whether stably inherited traits or epigenetic modifications affect gene regulation.

Keywords :
Bovine monocyte-derived macrophage, ex vivo infection, Mycobacterium avium ssp. paratuberculosis, RNA-sequencing, transcriptomics

P-01.12

ARE MERINOS MORE SUSCEPTIBLE TO OJD THAN CROSSBRED SHEEP? ANALYSIS OF ABATTOIR MONITORING FOR 2249 CONSIGNMENTS FROM THE NSW HIGH PREVALENCE AREA - 2003-2007.

Ian Links[1]

[1]Graham Centre for Agricultural Innovation, Wagga Wagga, NSW Australia 2650

IntroductionWhile there is strong anecdotal evidence that Merinos are more susceptible to OJD than Crossbred/British Breed flocks, there are few published reports in the literature ((Morris, Hickey et al. 2003). The results of abattoir monitoring were analysed for 2,249 direct consignments comprising 616,500 adult sheep (ﬄ2 years of age) from the NSW High Prevalence Area from 2003 to 2007 (Links, Denholm et al. 2007) where the breed category was recorded as Merino or Crossbred/British Breed at slaughter. ResultsMerinos comprised:2009/2249 of total consignments (av. 89.3%; annual range 85.9-98.3) and 578,508/616,500 (93.8%; range 91.9-99.2%) of sheep inspected.981/1036 (94.7%) of positive consignments327,654/340,940 (96.1%) of sheep in positive consignments.Consignments confirmed positive (ie from OJD infected properties):981/2009 (av. 48.8%; range 42.2-56.0%) of all Merino consignments.55/240 (av. 22.9%; range 14.0-27.5% excluding 2004) of all Crossbred consignments.Percentage of Sheep derived from Positive Consignments:56.6% (range 50.8-61.4%) of all Merino sheep monitored35.0% (range 25.9-37.6% excluding 2004) of all Crossbred sheep monitored. Sheep with Lesions attributable to OJD:6728/6904 (97.5%) were Merinos.Positive consignments (2003-2007) – lesions attributable to OJD:Av. 2.1% of Merino sheep, which declined progressively from 4.1% in 2003 to 1.2% in 2006 before rising to 1.6% in 2007.Av. 1.3% of Crossbred sheep, which declined from 1.6% in 2005 to 0.6% in 2007.All consignments monitored (2003-2007) - lesions attributable to OJD:Av. 1.2% of Merino sheep, which declined progressively from 2.1% in 2003 to 0.9% in 2007.Av. 0.5% of Crossbred sheep, which declined progressively from 1.9% in 2003 to 0.2% in 2007.Percentage of Lesions in Positive Consignments (heavily infected consignments):Merinos - 14.6% (143/981) with ﬄ5% lesions, 4.5% (44/981) with ﬄ10% lesions, and 0.2% (2/981) exceeding 30% lesions.Crossbreds - 3.6% (2/55) with ﬄ5% lesions, none exceeded 10% lesions.ConclusionData from abattoir monitoring in the NSW High Prevalence Area from 2003-2007, supports the hypothesis that Merinos (the dominant breed in the Australian sheep industry) are more susceptible to OJD than Crossbred sheep. The decline in prevalence from 2003 to 2007 is considered attributable to increasing levels of vaccination and implementation of Property Disease Management Programs.

Keywords :
Sheep, ojd, breed susceptibility, abattoir monitoring, Mycobacterium avium subsp. paratuberculosis, abattoir monitoring

P-01.13

MOLECULAR CHARACTERIZATION OF THE IMMUNE RESPONSE TO SHEEP NATURALLY INFECTED WITH MYCOBACTERIUM AVIUM SUBSP. PARATUBERCULOSIS VACCINATED WITH THE P35 PROTEIN EXPRESSED IN RECOMBINANT M13 PHAGE

González-Rodríguez D[1], Ramírez-Salcedo J[2], Manoutcharian K[3], **Chávez-Gris G**[1].

[1]UNAM, FMVZ, CEIEPAA, Tequisquiapan, Queretaro, Mexico. [2]UNAM, IFC, México3 UNAM, IIB, Mexico

Paratuberculosis is an infectious and chronic infection caused by acid alcohol resistant bacilli Mycobacterium avium subsp. paratuberculosis. Currently, there are different methods of control of paratuberculosis, including vaccination. The commercial vaccines help to prevent clinical disease and diminish clinical symptoms, but do not prevent infection; also interfere with the diagnosis and control of bovine tuberculosis. Currently, in Mexico there is no vaccine available. M. paratuberculosis has a membrane protein of 35 kDa (P35), immunodominant antigen that stimulates cellular immune response, which is reported in M. avium and M. leprae, but it isn´t present in M. tuberculosis or M. bovis. It's an immunodominant antigen which stimulates cellular immune response which is able to control the infection and limit the spread of bacteria to other cells. The aim of this study was to perform molecular characterization of the immune response through heterologous microarray, at 2 and 5 weeks post-vaccination of three sheep naturally infected with M. paratuberculosis vaccinated with the P35 protein expressed in recombinant M13 phage to identify the type of response generated by the vaccine. Differential gene expression and modification of various metabolic pathways was observed, but it is skewed to related to the immune response. At 2 and 5 weeks post-vaccination, changes in the gene expression were found in the Th1 response related with cell signaling pathways, intracellular traffic, adhesion and Th2 response. These results differ from those obtained in experimental infections previously reported. We conclude that the main change in expression in sheep vaccinated with recombinant P35 protein was genes of Th1 response, this response is related to the cellular immune response able to control the infection.ReferencesBannantine JP, Huntley JF, Miltner E, Stabel JR, Bermudez LE. The Mycobacterium avium subsp. paratuberculosis 35 kDa protein plays a role in invasion of bovine epithelial cells. Mycrobiol. 2003;149(8):2061-9.This project was supported by UNAM-DGAPA, PAPIIT IT202914Basagoudanavar SH, Goswami PP, Tiwari V. Cellular immune responses to 35 kDa recombinant antigen of Mycobacterium avium paratuberculosis. Vet Res Commun. 2006;30(4):357-67.

Keywords :
Paratuberculosis, P35, vaccination, microarrays

P-01.14

LOCAL IMMUNE RESPONSE ON LESIONS ASSOCIATED TO OVINE PARATUBERCULOSIS EVALUATED BY TISSUE MICROARRAYS

Chávez-Gris G[1], Vázquez-Arvízu S[1], Maldonado-Castro E[1], Castrellón-Ahumada V E[1], Baay G[2], Huerta S[2].

[1]UNAM, FMVZ, CEIEPAA, Tequisquiapan, Queretaro, Mexico. [2] Hospital Infantil de Mexico, Mexico

Cases diagnosed to ovine paratuberculosis 1998-2013 referred to Veterinary Faculty (Universidad Nacional Autonoma de Mexico) were selected. These lesions were classified into various types according to their severity, using stains of H-E and ZN. Subsequently was performed immunohistochemistry for detection of IL-10, IFN-gamma and INOs and were evaluated using tissue miroarrays to quantify the degree of expression of these markers in the different types of lessions, employing a program of Citometry (APERIO, Leica). These results showed that IFN-gamma is more evident in paucibacillary and less severe lesions. On the other hand, multibacillary are dominated by IL-10 with a decrease in the expression of IFN-gamma. ReferencesAlzuherri HM, Woodall CJ, Clarke CJ. Increased intestinal TNF-alpha, IL-1 beta and IL-6 expression in ovine paratuberculosis. Vet Immunol Immunopathol. 1996;49(4):331-45.

Keywords :
Immunohistochemestry, tissue microarrays, IFN gamma, IL10, histopatlogy

P-01.15

DYSBIOSIS OF THE FECAL MICROBIOTA IN CATTLE INFECTED WITH MYCOBACTERIUM AVIUM SUBSP. PARATUBERCULOSIS

Marie-Eve Fecteau[1], Dipti W. Pitta[1], Bonnie Vecchiarelli[1], Nagaraju Indugu[1], Sanjay Kumar[1], Susan C. Gallagher[1], Terry L. Fyock[1], **Raymond W. Sweeney**[1]

[1]*University of Pennsylvania School of Veterinary Medicine, Department of Clinical Studies-New Bolton Center, Kennett Square, PA 19348 USA*

The purpose of this study was to investigate the diversity patterns of fecal bacterial populations in cattle infected with MAP, compared to those of uninfected control cattle, using phylogenomic analysis. It is speculated that dysbiosis may contribute to the intestinal inflammation and diarrhea observed in cattle with Johne_disease. Fecal samples were tested from 20 MAP-positive but subclinical) cows; 25 MAP-negative herdmates; and 25 MAP-negative cows from a MAP-free herd. From each fecal sample the genomic DNA was extracted, 16S rDNA gene sequences PCR amplified and sequenced on a 454 Roche platform, and analyzed using QIIME. Approximately 199,077 reads were analyzed from 70 bacterial communities (average of 2,843 reads/sample). The composition of bacterial communities differed between the three treatment groups ($P < 0.001$; Permanova test). Taxonomic assignment of the OTUs identified 17 bacterial phyla across all samples. Bacteroidetes and Firmicutes constituted more than 95% of the bacterial population in the negative and exposed groups. In the positive group, lineages of Actinobacteria and Proteobacteria increased and those of Bacteroidetes and Firmicutes decreased ($P < 0.001$). Actinobacteria was highly abundant (30% of the total bacteria) in the positive group compared to exposed and negative groups (0.1 – 0.2 %). Notably, the genus Arthrobacter was found to predominate Actinobacteria only in the positive group. This study indicates that MAP-infected cattle have a different composition of their fecal microbiota than MAP-negative cattle.

Keywords :
Microbiome, dysbiosis

P-01.16

MYCOBACTERIUM AVIUM SUBSPECIES PARATUBERCULOSIS INFECTION PROGRESSION IN THE RABBIT MODEL

Molina E[1], **Arrazuria R**[1], Serrano M[1], Sevilla IA[1], Perez V[2], Garrido J[1], Juste RA[1,3], Elguezabal N[1]

[1]*Department of Animal Health, NEIKER-Instituto Vasco de Investigación y Desarrollo Agrario. Derio, Bizkaia, Spain.* [2]*Department of Animal Health, Faculty of Veterinary Medicine, University of León, Spain.* [3]*SERIDA, Department of Agriculture of the Regional Government of the Principality of Asturias, Deva, Asturias, Spain*

The mechanisms that operate behind Mycobacterium avium subsp. paratuberculosis (MAP) infection have probably not been fully elucidated due to the lengthy incubation period of the disease and the lack of an adequate and economic small animal model. In previous works we have shown that the rabbit MAP infection model may be useful. In order to better characterize this animal model this study aimed to assess infectious status at two different time points specially focusing on bacteriological and pathological aspects of the disease.New Zealand White rabbits were divided in two groups: healthy non-exposed controls (CNI, n=9) and infected controls orally challenged with MAP strain K10 (CI, n=9). Four animals from each group were necropsied at 75 days post-infection (dpi) and the remaining rabbits were necropsied at 165 dpi. The infection was assessed by qPCR, culture, PPA-3 ELISA and histopathology of gut associated lymphoid tissues.The CNI group did not present lesions compatible with PTB, neither was MAP detected by culture or qPCR. All of the animals from the CI group at 75 dpi presented at least one MAP positive tissue (sacculus rotundus or vermiform appendix) by culture and by qPCR and granulomatous lesions were present in 75% of the animals with a mean lesion severity index of 1. Animals belonging to the CI group that were necropsied at 165 dpi were not positive for MAP culture and only 40% were positive by MAP qPCR for at least one tissue (sacculus rotundus or mesenteric lymph node). However, 60% of these animals presented granulomatous lesions compatible with PTB with a mean lesion severity index of 1,71. ELISA readouts revealed no significant differences among groups independent of infectious status or necropsy date.In conclusion, in this short term rabbit infection model, MAP is eventually cleared from tissues and but lesion affected tissue types and lesion severity increase with time. Future studies should include more necropsy time points to verify if infection progression continues or on the contrary the increase in lesion severity is a reaction that leads to a more efficient elimination of MAP on the long term.

Keywords :
Mycobacterium avium subsp. paratuberculosis, rabbit, animal model

P-01.17
Identification of polymorphism in CLEC7A gene and their association with occurrence of Paratuberculosis in cattle

Satish Kumar[1], Subodh Kumar[1], Ran Vir Singh[1], Amit Kumar[1], Saurabh Gupta[2], Kundan Kumar Chaubey[2], Naveen Kumar[3], **Shoor Vir Singh**[2]

[1]*Division of Genetics, Indian Veterinary Research Institute, Izatnagar, Bareilly, India .* [2]*Microbiology laboratory, Animal Health Division, Central Institute for Research on Goats, Makhdoom, PO-Farah, Mathura, 281122, India.* [3]*Veterinary Type Culture Collection, National Research Centre on Equines, Hisar, Haryana-125001, India*

Paratuberculosis, caused by Mycobacterium avium subsp. paratuberculosis (MAP), is a chronic, infectious, granulomatous, inflammatory bowel disease primarily infecting domestic ruminants leading to persistent diarrhea, progressive wasting and eventually death. TLR4 has been implicated in cellular recognition of mycobacteria, binding cell wall components including lipoproteins. Bovine Dectin-1 is an important pattern recognition molecule that is able to generate a proinflammatory response by acting alongside Toll like receptor (TLR) signaling and is known to co-operate with TLR2 to specifically activate a macrophage proinflammatory response against mycobacterial infections.The present study was undertaken with aim to identify single nuceotide polymorphism in the gene encoding bovine Dectin-1 (CLEC7A) and to evaluate association of these SNPs with occurrence of paratuberculosis in cattle. For this 213 cattle belonging to four breed groups and four different farms were subjected to Johnin PPD, ELISA test (indigenous as well as Parachek kit method), faecal microscopy and faecal culture for detection of presence of bovine paratuberculosis infection. Based on the screening results 51 animals each could be assigned to case and control population. All the further investigations were done on these 102 animals. A total number of 6 SNPs viz rs110353594, rs110671821, rs110343521, rs41654445, rs109429379 and rs109280145 in CLEC7A gene were validated by amplifying five fragments and digested them with RE AlwNI, AluI, PsiI, NlaIV, Eco53kI and TaqI respectively. All six SNPs were found to be polymorphic in case: control population. The association study was carried out by PROC LOGISTIC procedure of SAS9.3. The SNP rs41654445 yielded three genotypes viz. CC, CT and TT with genotype frequencies in case and control population was 7.84, 41.18 and 50.98 and 66.67, 33.33 and 0 % respectively. These genotypes were significantly ($P<0.01$) different in case as compared to control cattle population. The ODDs of CC and CT verses TT genotype were <0.001(<0.001->999.99; 95 % CI) and <0.001(<0.001->999.99; 95 % CI) respectively which were significantly ($P<0.05$) lower among case population than control population. So TT genotype had more prevalence than CC and TT genotype in case population which is revealed by allele frequency of C and T in case and control population was 28.43 and 71.57 and 83.33 and 16.67 respectively.

Keywords : Paratuberculosis, Mycobacterium avium subsp. paratuberculosis, Toll like receptor, SNPs, Indigenous ELISA

P-01.18
Differences in the course of disease and lesions in goats experimentally infected with either Mycobacterium avium subsp. paratuberculosis (MAP) or Mycobacterium avium subsp. hominissuis (MAH)

E. M. Liebler-Tenorio[1], J. Schinköthe[1,2], P. Möbius[1], H. Köhler[1]

[1]*Institute of Molecular Pathogenesis, Friedrich-Loeffler Institut, Jena, Germany.* [2]*currently: Dept. of Experimental Animal Facilities and Biorisk Management, Friedrich-Loeffler Institut, Insel Riems, Germany*

Introduction: The closely related nontuberculous mycobacteria, MAP and MAH cause distinct symptoms and target different species. MAP is the etiologic agent of granulomatous enteritis and lymphadenitis in ruminants. MAH causes granulomatous lesions in the digestive tract of pigs and rarely other species, lymphadenitis in children and disseminated lesions in immunocompromised humans. The objective of this investigation was to compare the course of disease and lesions caused by MAP versus MAH in one host species, the goat.Material and methods: Goat kids were orally inoculated with MAP or MAH (10 times, every 2-3 days) receiving a total dose of 6.3x108 CFU MAP (strain 03A1961, bovine isolate) or 2.1x1010 CFU MAH (strain 09MA1289, porcine isolate). MAP-inoculated goats were necropsied 12 weeks after the last inoculation (wpi, n= 5) and 48 wpi (n=10); those inoculated with MAH 3-10 wpi (n= 9) and 48 wpi (n=9). Lesions were characterized in HE-stained sections. Mycobacteria were labeled in sections by immunohistochemistry and isolated by culture.Results: MAP-inoculated goatswere clinically healthy throughout the experiment, but lesions were detected in all of them. At 12 wpi, severe granulomatous infiltrates were seen in organized gut-associated lymphoid tissues (oGALT) especially in the jejunum (JPPs) and in intestinal lymph nodes (ILNs), and high amounts of MAP were isolated. At 48 wpi, atrophy of JPPs with reactive inflammatory infiltrates and granulomatous infiltrates and granulomas in ILNs predominated. Low to moderate amounts of MAP were isolated.Nine of the MAH-inoculated goats developed progressive disease. They died or were euthanized 3-10 wpi with severe ulcerative and granulomatous lesions especially in the ileal Peyer´s patch (IPP) and extensive necrosis in ILNs. Moderate to high amounts of MAH were isolated. Surviving goats had mild transient symptoms and were in good general condition at 48 wpi. Numerous large granulomas were detected in ILNs of all goats and small ones in IPP, but only low amounts of MAH.Conclusions: These differences reflect distinct host-pathogen interactions. Immune responses to MAH seem to be stronger which may be detrimental initially, but more efficient finally. MAP, by contrast, is better equipped for long-term coexistence with its host.

Keywords : Mycobacterium avium subsp. paratuberculosis, or Mycobacterium avium subsp. hominissuis, goat, pathology

P-01.19

Long term survival of animals positive to MAP diagnostic tests in Ireland

Peter Mullowney[1], Kevin Kenny[2], Margaret Good[1], John Egan[2]

[1]Department of Agriculture, Kildare Street, Dublin 2. [2] Central Veterinary Research Laboratory,, Department of Agriculture, Fisheries and Food Laboratories Complex, Backweston Campus, Stacummy Lane, Celbridge, Co. Kildare, Ireland

Objectives:To determine longevity of animals positive on a seroprevalence survey for Mycbacterium avium paratuberculosis and of animals positive on a faecal test submitted on clinical animals to the Regional Veterinary Laboratories.Materials and Methods: Three groups of animals were included in the survival analysis.1. Animals positive on the 2005 seroprevalence survey (Good et al., 2009)2. Animals positive on a faecal test submitted on clinical animals to the Regional Veterinary Laboratories in 2005.3. Two control animals were matched with each positive animal. These animals were picked from herds that had been tested negative on the prevalence survey and were matched on enterprise type, herd size and age of animal. Results:201 animals were positive on the prevalence survey. 8 were still alive eight years later. Of the 192 animals that were dead, one had been exported, 13% had died on the farm and 85% were sent for slaughter. Of the 400 control animals twenty five animals were still alive 8 years later. Of the 375 animals that were dead, three had been exported, 13% had died on the farm and 85% were sent for slaughter. Animals positive on a faecal test.20% had been sold. 20 were sent to the slaughterhouse, 18 died on the farm and one animal was exported. Animals survived from 1 to 1056 days after a positive faecal test. Of the 80 control animals 5.7% were still alive 8 years later. Of the 74 animals that were dead, one had been exported, 16% had died on the farm and 84% were sent for slaughter. Conclusions:The survival of positive animals on the prevalence survey compared with controls shows that on average positive animals died a year sooner than controls. Of the 40 animals positive on the faecal sample, the great majority were dead within 250 days of the sample being taken whereas 5.7% of the control animals were still alive five years later.

Keywords :
Longevity, fecal culture, antibody, Long term survival, positive to MAP diagnostic tests.

P-01.20

Monitoring paratuberculosis status in dairy cattle in Minnesota, USA: A descriptive analysis based on a voluntary screening program

Kanankege Kaushi[1], **Picasso Catalina**[1], Dokkebakken Bruce[2], Schumann Valerie[2], Wells Scott J.[1], Perez Andres M.[1], Alvarez Julio[1]

[1]Veterinary Population Medicine, University of Minnesota, Saint Paul, MN, USA. [2]Minnesota Dairy Herd Improvement Association, MN, USA.

Mycobacterium avium subsp. paratuberculosis (MAP) is the causative agent of paratuberculosis (PTB), a chronic debilitating enteritis affecting cattle, sheep and goats. Minnesota (MN) has 460,000 dairy cattle and is 7th in milk cows in the U.S. PTB control is not legislated in the U.S. (or MN), which results in limited estimates of disease burden and focusing disease control on voluntary producer-driven programs. Here, we describe the epidemiological features of data collected through the Minnesota Dairy Herd Improvement Association (Minnesota DHIA) on MAP screening using a milk ELISA, from November 2013 to November 2014. This study was aimed at characterizing disease patterns and assessing economic losses, to ultimately elaborate recommendations for an effective voluntary PTB control program in MN. The Minnesota DHIA database represents nearly 1,800 (50%) dairy herds in MN. Data assessed here includes a total of 35,492 ELISA milk screening assays performed on 32,672 unique cows belonging to 633 herds between November 2014 and November 2015 were classified as positive (P), negative (N), and inconclusive (S). Herd and individual characteristics were compared based on the test results. Number of herds tested each month ranged between 1 and 312, with monthly number of herd positivity (i.e., herds with at least one ELISA positive animal) falling consistently within 0-50 range. Distribution of positive herds across the state was heterogeneous. At the individual level, several production variables were associated with increased odds of testing positive, some of which were also related with the age of the animal. These preliminary results suggest that results from the voluntary screening program may be useful to assess the situation of PTB at the state level. In addition, they may help to quantify production losses at the farm and individual level associated with the presence of disease.

Keywords :
Monitoring, paratuberculosis, dairy cattle, voluntary program, milk ELISA

P-01.21
IMMUNOLOGICAL NATURE OF SHEEP INHERENTLY RESISTANT TO MAP INFECTION

Kumudika de Silva[1], Douglas Begg[1], Auriol Purdie[1], Karren Plain[1], Richard Whittington[1]

[1]*School of Life and Environmental Sciences, Faculty of Veterinary Science, The University of Sydney, Australia*

Understanding the nature of the protective immune response which shields an animal from the development of Johne's disease is necessary for developing an effective vaccine. Deliberate manipulation of the host immune response towards a similar profile should improve vaccine-induced protection against MAP infection. In this unique long-term longitudinal study a cohort of 20 sheep experimentally exposed to MAP and 10 controls were monitored for 2.5 years allowing for the slow progression of Johne's disease under 'natural' conditions. Cellular and humoral immune parameters and faecal shedding were examined for the duration of the trial and disease outcomes were classified based on the presence of viable bacteria and pathological lesions in intestinal tissues at the termination of the trial or when animals were culled due to weight loss. Forty percent of MAP-exposed sheep were removed from the trial due to mild weight loss at 1-2 years post MAP exposure. Of the animals that survived, one had severe pathology which was detected only by post mortem tests. The majority (73%) of MAP-exposed disease-free (no viable MAP detected by culture and no histological lesions in 6 intestinal tissue and associated lymph node sections) sheep were infectious (MAP detected in faeces in at least one sampling), shedding MAP intermittently, prior to the removal of animals with clinical disease. Similar to diseased (viable MAP detected by tissue culture) sheep, the immune responses of disease-free animals were dynamic. Immune response patterns of infectious and non-infectious disease-free sheep will be discussed. This study confirms previous work in a shorter trial which demonstrated that early immune responses reflect eventual disease outcome (de Silva et al 2013). Identification of animals likely to be inherently resistant to MAP infection may also be useful to inform strategies used to control the spread of disease. Reference : de Silva et al 2013 Prev Vet Med 112 : 203– 212

Keywords :
Immune response, resistance

P-01.22
EXPERIMENTAL INOCULATION OF DIFFERENT BREEDS OF SHEEP WITH MYCOBACTERIUM AVIUM SUB SPECIES PARATUBERCULOSIS; ARE BREED SUSCEPTIBILITY DIFFERENCES EVIDENT?

Douglas Begg[1], A.C Purdie[1], K. de Silva[1], K.M Plain[1], R.J Whittington[1]

[1]*School of Life and Environmental Sciences, Faculty of Veterinary Science, The University of Sydney, Camden, NSW, Australia*

Infection with Mycobacterium avium subspecies paratuberculosis (MAP) does not always lead to Johne's disease. Understanding why some animals enter a path to disease susceptibility or resilience is an important key to controlling mycobacterial diseases. As an aid to enable greater understanding of this issue, a study was designed to examine the susceptibility or resistance of various breeds of sheep to infection with MAP. Four breeds of sheep; Merino, Suffolk first cross Merino, Border Leicester, and Poll Dorset were orally inoculated with MAP and monitored for 14 months. Clinical disease rates were greater in the Merino (44%) and Suffolk first cross Merino breeds (37%) and lower for the Border Leicester (12%) and Poll Dorset (11%) breeds. Infection rates, as determined by culture of gut associated tissues, ranged from 75% for the Suffolk first cross Merino to 44% for the Poll Dorset sheep. There was no difference in the infectiousness via MAP faecal shedding of the clinical animals of any breed. Susceptibility to MAP infection, as determined by infection and clinical disease development, was observed in all the breeds examined in this study although there were differences in the severity of disease between the breeds. Poll Dorset and Border Leicester sheep appeared more resilient to MAP infection but there was evidence that more animals would have of these breeds were likely to develop disease if the duration of the trial had been extended. These findings provide evidence of a potential genetic association with the rate of disease progression and are important in the understanding of disease pathogenesis, as well as in the identification of the risks of disease spread, and may have an influence on control programs for MAP.

Keywords :
Sheep, breed, pathology

P-01.23
BIPHASIC MAP FAECAL SHEDDING PROFILE OVER TIME IN AN EXPERIMENTAL CATTLE INFECTION MODEL

Karren Plain[1], Douglas J. Begg[1], Kumudika de Silva[1], Auriol C. Purdie[1], Richard J. Whittington[1]

[1]*School of Life and Environmental Sciences, Faculty of Veterinary Science, The University of Sydney, Australia*

Experimental models of Johne's disease in the natural host are logistically challenging due to the need for long term monitoring. This is particularly the case in cattle, where the disease can take many years to manifest. In one of the largest and longest studies of its kind, an experimental infection trial was conducted in 20 exposed and 10 control cattle monitored intensively over a period of nearly 5 years. A low inoculation dose of C strain MAP was given orally to calves; this was intentional to ensure that the outcomes would be comparable to natural infection. Frequent (1-3 monthly) blood and faecal sampling was used to monitor immune and infection parameters, and intestinal biopsies were performed at two time points during the subclinical disease phase. Faecal shedding was monitored using a sensitive liquid culture method and a validated faecal PCR test. Although clinical disease was not seen, there was evidence of infection in 37% of the animals and 10% had histopathological lesions consistent with Johne's disease. The latter is similar to the proportion that would be expected to succumb to clinical disease in natural infection. Faecal shedding occurred in two distinct phases; intermittent early shedding <6 months post-exposure that did not correlate to disease outcomes, followed by a smaller cohort of animals that progressed to more consistent shedding in the later stages of the trial. The infected cattle could be grouped into low and high IFN-gamma responders based on their response at the 4 month time point, with mean IFN-gamma responses of the animals in the high group remaining elevated compared to the low IFN-gamma responder group, which comprised the animals that ultimately developed disease. The biopsy specimens provided evidence of regression of a lesion in the intestine of one of the infected cattle, a phenomenon that has previously been reported to occur in sheep with Johne's disease. The information from this trial has led to a greater understanding of the changes that occur during the disease course in cattle.

Keywords :
Faecal sheeding, experimental infection model, cattle, immune response

P-01.24
MYCOBACTERIUM AVIUM SSP. PARATUBERCULOSIS SHEDDING PATTERNS IN QUÉBEC DAIRY COWS

Juan Carlos Arango Sabogal[1], Julie Paré[2], Olivia Labrecque[3], Jean-Philippe Roy[1], Geneviève Côté[4], Sébastien Buczinski[1], Vincent Wellemans[1], Gilles Fecteau[1]

[1]*Département de sciences cliniques, Faculté de médecine vétérinaire, Université de Montréal, Saint-Hyacinthe, Québec, Canada.* [2]*Agence canadienne d'inspection des aliments, Saint-Hyacinthe, Québec, Canada.* [3]*Laboratoire d'épidémiosurveillance animale du Québec, Ministère de l'Agriculture, des Pêcheries et de l'Alimentation du Québec, Saint-Hyacinthe, Québec, Canada.* [4]*Direction générale des laboratoires d'expertise, Ministère de l'Agriculture, des Pêcheries et de l'Alimentation du Québec, Québec, Canada*

Mycobacterium avium ssp. paratuberculosis (MAP) infection has been divided in 4 stages according to severity of clinical signs, shedding level and ease of detection. Progression through these stages depends, among others, on age at initial exposure and initial MAP-dose. Most cows will not progress to clinical stage. Some will contain infection while others will be culled before. Most naturally infected animals show a low and intermittent shedding pattern and never progress to high shedding.This study describes the shedding patterns of cows identified as positive by individual fecal culture (IFC) over a 4 years period.Data of 27 Holstein, 16 Jersey and 6 Ayrshire cows from 21 Québec dairy herds were purposively selected from existing databases. Cows were IFC positive at least once during the annual herd sampling (2010-2014). MAP was isolated using the liquid media in an automated detection system.Cows were tested at least once and a maximum of 5 times during the study period. Mean age at first positive result (FPR) was 4.8 years (range: 2.1 to 9.4 years). Mean time between FPR and culling was 1 year (range: 0.1 to 4.3). Paratuberculosis positive result was the most common cause of culling (n=23), followed by inadequate milk production (n=6), reproduction failure (n=3), musculoskeletal problem (n=2), sudden death (n=2), bovine leukosis (n=1), and poor body condition (n=1). Reason for culling was not available for 11 cows. Most cows were culled after their FPR (n=36). Four cows followed a shedding pattern suggestive of intermittent excretion. They were sampled 3 to 5 times over a 3 years period. Nine cows remained in the herd on average 1.7 years (range: 8 months to 2.9 years) even after 1 or more positive results.Results suggest evidence of intermittent fecal excretion of MAP in Québec dairy cows. Changes in culture tests results of non-progressor cows may be the consequence of variation in test sensitivity, intermittent shedding or pass-through phenomenon. Culling after FPR was observed in most of the cows; however some progressor animals were kept in the herds even after several positive results indicating that other culling reasons may prevail over MAP status.

Keywords :
Mycobacterium avium ssp. paratuberculosis, shedding patterns, intermittent shedding

P-01.25

Histopathological characteristics in new experimental autoimmune encephalomyelitis (EAE) induced by Mycobacterium avium subsp. paratuberculosis in mice

Eiichi Momotani[1], Shigetoshi Eda[2], Davide Cossu[1,3]

[1]Tohto College of Health Sciences, Tsukuba, Japan. [2]Department of Forestry, Wildlife and Fisheries, the University of Tennessee, USA. [3]current affiliation: Department of Neurology, Juntendo University School of Medicine, Japan

The experimental autoimmune encephalomyelitis (EAE) has been shown to be a useful model of the human multiple sclerosis (MS). This model is organ specific and appears to be primarily a cell-mediated disorder similar to the acute form of EAE, however the differences have been discussed. One of the big issue to extrapolate the model to explain the pathogenesis of real disease is using Freund_omplete adjuvant containing Mycobacterium tuberculosis(MTB). Mycobacterium avium subsp. paratuberculosis (MAP) has been suspected to be the etiological agent of human autoimmune diseases, such as Crohn's disease, Type 1 Diabetes, Multiple Sclerosis and Hashimoto's Thyroiditis. We have reported CD model mouse by using MAP antigen in mouse, however there is no available experimental model with MAP for other diseases yet. Since Th1 and Th2 response for MAP antigen were previously reported, we made mouse EAE model by using MAP antigen. We replaced MTB used in traditional EAE model, with MAP. Same concentration of MAP antigen to the traditional one was mixed with Incomplete Freund's adjuvant (IFA) and MOG35-55 synthesized peptide was emulsified by ultra-sonication. C57BL/6 mouse were subcutaneously inoculated. Pertussis toxin was injected according to standard EAE method. Positive control of original EAE mouse and negative control mouse injected inoculum without mycobacterial antigens were used. Mice were observed dairy and finally sacrificed with overdose anesthesia and sampled for histopathology. Clinical symptom was recorded as following grade. Grade 1, tail weakness ; 2, mild paralysis and/or ataxia of the hind limbs; 3, severe paralysis of the hind limbs; 4, moribund; 5, death due to EAE. First clinical sign of grade 1 was observed 9th day of the injection, some mice showed grade 2 in 10 days and 3 after 11 days. Date of occurrence was similar in traditional and new MAP-EAE, however the intensity was more severe in MAP one. Histopathological examination revealed perivascular cuffing with lymphocyte/macrophage, demyelination, reactive gliosis, swollen axon sheaths, in the meninges and in the white matter. These studies reveled first pathological suggestion including common antigenicity and/or adjuvant actibity in the contribution of MAP antigen in MS.

Keywords :
MS, EAE, pathology, Experimental, mouse, autoimmune, model

P-01.26

Association between genetic polymorphisms and epidemio-pathogenic forms of paratuberculosis

Juste RA[1,2], Vazquez P[1,3], Ruiz-Larrañaga O[4], Iriondo M[4], Manzano C[4], Agirre M[4], Estonba A[4], Geijo MV[1], Molina E[1], Sevilla I[1], Alonso-Hearn M[1], Gomez N[1], Perez V[4], Cortes A[6], Garrido JM[1]

[1]NEIKER-Tecnalia, Basque Institute for Agriculture Research and Development, Derio, Bizkaia, Spain. [2]SERIDA, Regional Service of Agricultural Research and Development, Villaviciosa, Asturias, Spain. [3]SALUVET, Department of Animal Health, Faculty of Veterinary Sciences. Complutense University of Madrid (CUM). [4]Department of Genetics, Physical Anthropology and Animal Physiology, Faculty of Science and Technology, University of the Basque Country (UPV/EHU), Leioa, Bizkaia, Spain. [5]Department of Animal Health. Veterinary School. University of Leon. Leon, Spain. [6]Department of Health. Basque Government, San Sebastian – Donostia. Gipuzkoa, Spain.

Paratuberculosis control is a challenging issue that requires different approaches. Use of genetic markers for improving resistance to infection has been explored as a promising population strategy, but up to now no useful markers have been identified. In a study where 502 slaughtered Friesian cows were submitted to histopathological and genetic study involving 24 SNPs in six candidate genes (NOD2, CD209, SLC11A1, SP110, TLR2 and TLR4), the SNPs with more extreme frequencies were selected for testing the association of combinations of 2, 3, 4, 5 and 6 SNPs to paratuberculosis epidemio-pathogenic (EP) forms (apparently free-AF, latent-LAT and patent-PAT). Best EP form grouping was obtained using a combination of 5 SNPs in 4 genes (CD209, SLC11A1, SP110 and TLR2). These groups were defined as paratuberculosis infection risk and showed the following distributions: low infection progression risk (LOWIN) with 39 (8%) cases (94.9% AF / 5.1% LAT / 0% PAT); infection latent risk (LATIN) with 17 (3%) cases (5.9% AF / 94.1% LAT / 0% PAT); average progression risk (AVERIN) with 413 (82%) cases (52.1% AF / 38.5% LAT / 9.4% PAT) and patent infection progression risk (PATIN) with 33 (7%) cases (36.4% AF / 24.2% LAT / 39.4% PAT. Age of slaughter was significantly higher for LATIN (88.3 months) compared to AVERIN (65.3 months; p=0.0007) and PROGIN (59.1 months; p=0.0004) and for LOWIN (73.9 months) compared to PROGIN (p=0.0233) and nearly significant compared to AVERIN (p=0.0572). These results suggest that some selected genetic polymorphisms have a potential to be used as markers of paratuberculosis EP forms and thus add a new tool for the control of this infection.

Keywords :
Paratuberculosis, genetic resistance, pathology, lifespan, cattle

P-01.27

EXPERIMENTAL EVALUATION OF THE EFFECTIVENESS OF ZEOLITE CLINOPTILOLITE FOR MYCOBACTERIUM AVIUM SUBSP. PARATUBERCULOSIS (MAP) CONTROL IN NATURAL CONTAMINATED DAIRY CATTLE SLURRY

Avilez Carolina[1,2], Alfaro Marta[4], Salazar Francisco[4], Encina Carolina[1], Verdugo Cristobal[1], Martínez Oscar[3], Collins Michael[5], Steuer Pamela[1], Tejeda Carlos[1], **Salgado, Miguel**[1]

[1]Preventive Veterinary Medicine Department, Faculty of Veterinary Sciences; Universidad Austral de Chile, Valdivia, Chile. [2]Graduate School, Faculty Veterinary Science; Universidad Austral de Chile, Valdivia, Chile. [3]Biochemistry and Microbiology Department, Faculty of Sciences; Universidad Austral de Chile, Valdivia, Chile. [4]Institute for Agricultural Research (INIA), Remehue Research Centre, Osorno, Chile. [5]Department of Pathobiological Sciences, School of Veterinary Medicine, University of Wisconsin, Madison, USA.

The use of slurry as fertilizer on grassland could efficiently spread infectious agents. Among those using fecal–oral way of transmission, Mycobacterium avium subsp. paratuberculosis (MAP) represents an interesting pathogen model to be studied in slurry management systems. A wide range of treatment processes to control pathogens already exists, but many of the systems are considered too expensive for livestock farming. In order to fulfill the requirement of a practical control implementation, we aimed to investigate the effect of natural zeolite, a mineral that has shown pathogen control properties, on MAP viability and concentration in dairy cattle natural contaminated slurry.
MAP viability and concentration in dairy cattle slurry after a separation physical treatment was estimated. A randomized block design constituted by four separation phases of treated slurry (liquid supernatant, solid supernatant, zeolite with slurry, liquid residue) plus one control (not filtered) was used. To assess MAP detection, each obtained sample was evaluated by culture sampling at different separation phases, along with MAP quantification directly from the same samples. Control cattle slurry samples showed a lower MAP concentration than the solid supernatant separation phase. Surprisingly, zeolite+slurry treatment showed the lowest number of viable MAP. The zeolite-based treatments reduced significantly the survival of MAP in treated slurry. The results suggest that zeolite treatments using the zeolite filter may be an interesting alternative of MAP control in slurry. The implementation of this treatment seems to be relatively simple and low cost. The use of zeolite treatments to control MAP and maybe other pathogens seems promising, however further research is need to understand and clarify the involved mechanisms that explain in detail the observed results.

Keywords :
Cattle slurry, MAP, zeolite, filter, control

NOTES

NOTES

NOTES

NOTES

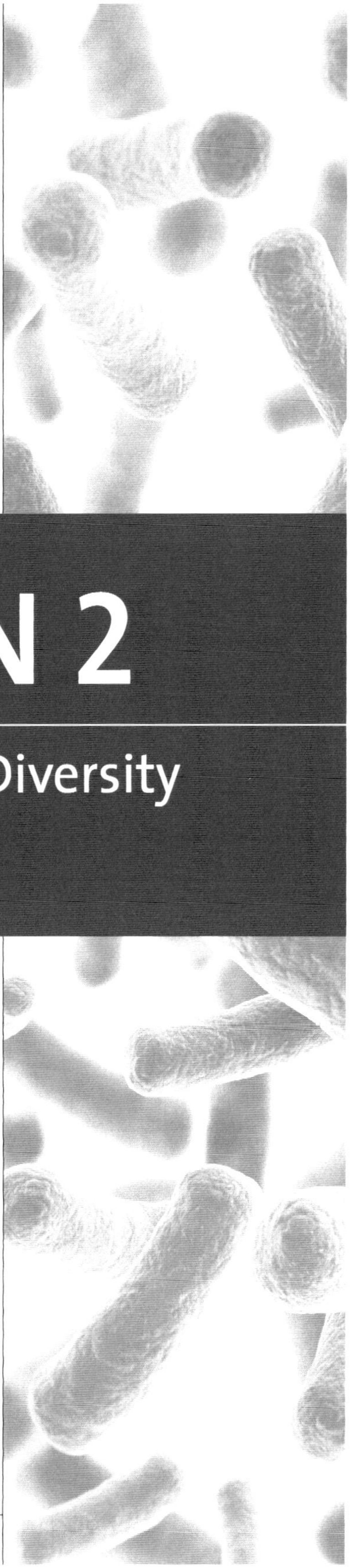

SESSION 2

Map genomics & Diversity

Tuberculosis and One Health: comparative analyses of human and bovine tubercle bacilli

Stephen V. Gordon, UCD School of Veterinary Medicine, University College Dublin, Ireland

Theobald Smith first differentiated human and bovine tubercle bacilli in 1896 [1]. In his seminal work he recognized the potential for comparative analyses of human and animal TB to catalyse the development of new control tools for TB, contending that comparative studies would "*lead eventually to more light on the whole subject of tuberculosis from the preventive as well as the therapeutic side*". However, the integration of research in human and animal TB has not been exploited to its full potential. This is a lost opportunity, as TB is an ideal paradigm for the One Health agenda and the interaction of human, animal and environmental health. In this presentation I will discuss work on comparative analyses at the genomic, transcriptomic and proteomic level of *Mycobacterium tuberculosis and Mycobacterium bovis* as the exemplar human- and animal-adapted TB complex strains. I will describe what these studies have revealed about the virulence, evolution and host adaptation of the tubercle bacilli, and hopefully offer insight into how similar approaches could be pursued in the study of *Mycobacterium avium* subsp *paratuberculosis* (MAP).

Observations of the distinct host preferences of tubercle bacilli date back to the work of Koch, Von Behring, Smith and others. In seminal work Theobald Smith examined 8 human isolates (one isolated from a pet of a TB patient), 6 isolates from cattle, and single isolates from a pig, cat, and horse for their microscopic appearance, cultural characteristics, and pathogenesis in mice, guinea pigs, rabbits, pigeons and cattle [2]. The bovine isolates "*grew less vigorously for a number of generations*" in coagulated serum, while "*bovine bacilli tend to remain short; human bacilli are either more slender from the start or become so during cultivation*". In terns of pathogens, bovine bacilli had "*a much greater pathogenic activity towards rabbits, guinea- pigs and cattle*" than human isolates. This work defined the bovine tubercle bacilli as distinct from the human isolates, and furthermore defined the distinctive virulence of the isolates in animal models [2].

These classic experiments of Smith on host preference were revisited using well characterised *Mycobacterium tuberculosis* complex (MTBC) bacilli, namely the genome sequenced strains *M. tuberculosis* H37Rv and *M. bovis* AF2122 [3]. Cattle infected with *M. tuberculosis* H37Rv and *M. bovis* AF2122 became positive to skin-test and interferon gamma release assays; however, the *M. tuberculosis* H37Rv infected cattle showed no pathological signs of disease (even though same *M. tuberculosis* H37Rv seed lot used to infect cattle caused disease in guinea pigs). Although these experimentally infected cattle did not show obvious pathological signs, cattle naturally infected with *M. tuberculosis* that show typical TB granulomatous lesions and are culture positive for *M. tuberculosis* have been identified in multiple studies in regions such as Ethiopia, Nigeria, or China [4-6]. One can however argue that rather than *M. tuberculosis* sustaining in these cattle populations, they instead represent reverse zoonotic infections in countries where the burden of human TB disease is high and immune status of cattle is compromised [7, 8]. One can argue therefore that cattle infected with *M. tuberculosis* recapitulate key presentations of TB in humans, from latent infection to active disease. The bovine-*M. tuberculosis* infection model may therefore present a unique model in defining the *M. tuberculosis*-host dynamic.

The key to a pathogen's life cycle is the possession of virulence systems that enable it sustain in a host population. These virulence factors run the gamut from battering rams that act across multiple hosts to lock-picks that open a host-specific backdoor. Defining virulence factors involved in host adaptation is complex, not least because 'virulence' depends on context; the host's immune status and genetic background have a major affect on the outcome of infection [9]. While great strides have been made in defining virulence factors in the MTBC, these have largely been defined on the interaction between *M. tuberculosis* mutants and murine infection models. However the focus on *M. tuberculosis* and the murine model is not without its difficulties. For example, the RD1 region was originally identified as a region deleted from *M. bovis* BCG relative to virulent *M. bovis* or *M. tuberculosis* [10]; inactivation of ESX-1 machinery or ESX-1-secreted effectors attenuate *M. tuberculosis* and *M. bovis* in animal models [11, 12]. Hence ESX-1 can fairly be described as a locus that is essential for virulence in these strains and models. However, M. mungi, M. microti, or the Dassie bacillus all have RD1-like regions deleted, so ESX-1 systems are clearly not required for these strains to sustain in their respective host populations [13-15]. Hence, the presence of an intact ESX-1 locus is not sufficient for virulence.

PhoPR is a two component regulatory system consisting of the PhoP and PhoR components [16, 17]. It has been shown to play a key role in regulating the virulence of *M. tuberculosis*, where it is involved in the regulation of genes for the synthesis of several lipids and secretion of ESAT-6. Guilhot and colleagues explored the role of SNPs in PhoPR across MTBC members, and showed that mutations in *M. africanum* and *M. bovis* led to reduced expression of the PhoPR regulon in these species [18]. When the same PhoP mutations were introduced into *M. tuberculosis* the recombinant strains had reduced virulence in both human macrophages in vitro and a mouse model of infection. Strikingly, the authors also showed that the RD8 deletion in *M. bovis* helps to compensate for the loss of PhoPR regulation to maintain secretion of ESAT-6 and ESX-1 substrates. Furthermore, an *M. bovis* strain that was isolated from a human-human transmission chain in a HIV ward was shown to posses an insertion sequence upstream of phoPR that drives its expression and hence compensates for the PhoPR mutations [18]. Hence PhoPR mutations could explain why, despite its high degree of genetic similarity to *M. tuberculosis*, *M. bovis* cannot sustain infection in human populations. Further study into emergence of this genotype should identify what adaptations were made in *M. bovis* to enhance its ability to maintain virulence strains in alternate animal hosts.

Two proteins that show differential expression across the MTBC are MPT70 and MPT83 (aka MPB70 and MPB83 in *M. bovis*). MPT83 is a lipoprotein that is post-translationally glycosylated, while MPB70 is secreted with no post-translational modifications [19]. *Mycobacterium bovis* shows constitutive high-level expression of these antigens, while *M. tuberculosis* has low-level expression but shows induction during intracellular growth [20]. Expression of MPB70 and MPB83 is under the control of the SigK regulon, with high-level expression in *M. bovis* resulting from a loss of negative regulation due to a mutation in the gene encoding the anti-sigma factor RskA [21]. Intriguingly, constitutive high expression of MPB70 and MPB83 is also seen in *M. orygis* by an independent missense mutation in rskA to that seen in *M. bovis*. The upregulation of the SigK regulon in *M. bovis* and *M. orygis* through independent mutations would suggest a selective advantage for increased expression of MPB83, MPB70 and other constituents of the regulon; however the nature of this advantage is unclear.

The phenolic glycolipids (PGL) of the MTBC have known roles in virulence and immune modulation. These glycolpids are built on a core of phthiocerol dimycoserosate (DIM), with members of the MTBC having variation in the carbohydrate structures that are linked to DIM [22]. *M. bovis* produces mycoside B, a monosaccharide variant with 2-O-methylrhamnose as the terminal sugar, a structure also seen in *M. microti* and *M. pinnipedii* [23]. A minority of *M. tuberculosis* strains produce a trisaccharide variant, with mutations in an assortment of glycosyltransferaes, methyltransfrases and polyketide synthases responsible for the variable production across *M. tuberculosis* lineages. Hence variation in PGL structures across the MTBC is apparent, but this variation occurs in the context of many other genetic differences, confounding simple linkages between lipid presence/absence to virulence. To address these problems genetic approaches have been used to reprogramme the synthesis of PGL across the MTBC allowing the impact of lipid modifications to be assessed in isogenic backgrounds; for example, to switch the mycoside B variant of *M. bovis* BCG to the PGL variant expressed by *M. leprae* [24]. Added to this, complete chemical synthesis of the PGL-tb has been recently reported, allowing the activity of this glycolipid to be studied in isolation [25]. Full chemical synthesis of the para-hydroxybenzoic acid derivatives (pHBADs), that contain an identical glycosylated phenolic moiety to PGLs, has been reported and have been used to show that pHBAD variants in isolation can suppress the production of IFN-γ and IL-17 by stimulated murine splenocytes [26]. Teasing apart the roles of PGL and pHBADs variants on host interaction across the MTBC now appears feasible.

The MTBC represent the ideal group of pathogens to explore concepts in One Health, with collaboration across human, veterinary and environmental spheres offering new insights into pathogen evolution, virulence, and disease transmission. With advances in genome sequencing technologies, and cognisant of undiscovered animal-adapted strains lying in waiting, we are now poised to look afresh at the diversity across the MTBC and the nature of host adaptation and virulence. Furthermore, the expansion of these approaches to the study of MAP, and interaction between the TB and ParaTB fields, offer the potential for new insight into the evolution of virulence and host tropism across mycobacterial pathogens in general.

1. *T S. Two varities of the tubercle bacillus from mammals. Trans Assoc Amer Physicians. 1896;11:75-95.*
2. *Smith T. A Comparative Study of Bovine Tubercle Bacilli and of Human Bacilli from Sputum. The Journal of experimental medicine. 1898;3(4-5):451-511. PubMed PMID: 19866892; PubMed Central PMCID: PMC2117982.*
3. *Whelan AO, Coad M, Cockle PJ, Hewinson G, Vordermeier M, Gordon SV. Revisiting host preference in the Mycobacterium tuberculosis complex: experimental infection shows M. tuberculosis H37Rv to be avirulent in cattle. PloS one. 2010;5(1):e8527. doi: 10.1371/journal.pone.0008527. PubMed PMID: 20049086; PubMed Central PMCID: PMC2795854.*
4. *Berg S, Firdessa R, Habtamu M, Gadisa E, Mengistu A, Yamuah L, et al. The burden of mycobacterial disease in ethiopian cattle: implications for public health. PloS one. 2009;4(4):e5068. doi: 10.1371/journal.pone.0005068. PubMed PMID: 19352493; PubMed Central PMCID: PMC2662418.*
5. *Cadmus S, Palmer S, Okker M, Dale J, Gover K, Smith N, et al. Molecular analysis of human and bovine tubercle bacilli from a local setting in Nigeria. Journal of clinical microbiology. 2006;44(1):29-34. doi: 10.1128/JCM.44.1.29-34.2006. PubMed PMID: 16390943; PubMed Central PMCID: PMC1351927.*
6. *Chen Y, Chao Y, Deng Q, Liu T, Xiang J, Chen J, et al. Potential challenges to the Stop TB Plan for humans in China; cattle maintain M. bovis and M. tuberculosis. Tuberculosis. 2009;89(1):95-100. doi: 10.1016/j.tube.2008.07.003. PubMed PMID: 19056318.*
7. *Ameni G, Vordermeier M, Firdessa R, Aseffa A, Hewinson G, Gordon SV, et al. Mycobacterium tuberculosis infection in grazing cattle in central Ethiopia. Veterinary journal. 2011;188(3):359-61. doi: 10.1016/j.tvjl.2010.05.005. PubMed PMID: 20965132; PubMed Central PMCID: PMC3103825.*
8. *Ocepek M, Pate M, Zolnir-Dovc M, Poljak M. Transmission of Mycobacterium tuberculosis from human to cattle. Journal of clinical microbiology. 2005;43(7):3555-7. doi: 10.1128/JCM.43.7.3555-3557.2005. PubMed PMID: 16000505; PubMed Central PMCID: PMC1169140.*
9. *Casadevall A, Pirofski LA. Host-pathogen interactions: basic concepts of microbial commensalism, colonization, infection, and disease. Infection and immunity. 2000;68(12):6511-8. PubMed PMID: 11083759; PubMed Central PMCID: PMC97744.*

10. Mahairas GG, Sabo PJ, Hickey MJ, Singh DC, Stover CK. Molecular analysis of genetic differences between Mycobacterium bovis BCG and virulent M. bovis. Journal of bacteriology. 1996;178(5):1274-82. PubMed PMID: 8631702; PubMed Central PMCID: PMC177799.

11. Lewis KN, Liao R, Guinn KM, Hickey MJ, Smith S, Behr MA, et al. Deletion of RD1 from Mycobacterium tuberculosis mimics bacille Calmette-Guerin attenuation. The Journal of infectious diseases. 2003;187(1):117-23. doi: 10.1086/345862. PubMed PMID: 12508154; PubMed Central PMCID: PMC1458498.

12. Wards BJ, de Lisle GW, Collins DM. An esat6 knockout mutant of Mycobacterium bovis produced by homologous recombination will contribute to the development of a live tuberculosis vaccine. Tubercle and lung disease : the official journal of the International Union against Tuberculosis and Lung Disease. 2000;80(4-5):185-9. doi: 10.1054/tuld.2000.0244. PubMed PMID: 11052907.

13. Alexander KA, Laver PN, Michel AL, Williams M, van Helden PD, Warren RM, et al. Novel Mycobacterium tuberculosis complex pathogen, M. mungi. Emerging infectious diseases. 2010;16(8):1296-9. doi: 10.3201/eid1608.100314. PubMed PMID: 20678329; PubMed Central PMCID: PMC3298296.

14. Brodin P, Eiglmeier K, Marmiesse M, Billault A, Garnier T, Niemann S, et al. Bacterial artificial chromosome-based comparative genomic analysis identifies Mycobacterium microti as a natural ESAT-6 deletion mutant. Infection and immunity. 2002;70(10):5568-78. PubMed PMID: 12228284; PubMed Central PMCID: PMC128332.

15. Mostowy S, Cousins D, Behr MA. Genomic interrogation of the dassie bacillus reveals it as a unique RD1 mutant within the Mycobacterium tuberculosis complex. Journal of bacteriology. 2004;186(1):104-9. PubMed PMID: 14679230; PubMed Central PMCID: PMC303463.

16. Perez E, Samper S, Bordas Y, Guilhot C, Gicquel B, Martin C. An essential role for phoP in Mycobacterium tuberculosis virulence. Mol Microbiol. 2001;41(1):179-87. PubMed PMID: 11454210.

17. Lee JS, Krause R, Schreiber J, Mollenkopf HJ, Kowall J, Stein R, et al. Mutation in the transcriptional regulator PhoP contributes to avirulence of Mycobacterium tuberculosis H37Ra strain. Cell Host Microbe. 2008;3(2):97-103. doi: 10.1016/j.chom.2008.01.002. PubMed PMID: 18312844.

18. Gonzalo-Asensio J, Malaga W, Pawlik A, Astarie-Dequeker C, Passemar C, Moreau F, et al. Evolutionary history of tuberculosis shaped by conserved mutations in the PhoPR virulence regulator. Proc Natl Acad Sci U S A. 2014;111(31):11491-6. doi: 10.1073/pnas.1406693111. PubMed PMID: 25049399; PubMed Central PMCID: PMCPMC4128152.

19. Wiker HG, Nagai S, Hewinson RG, Russell WP, Harboe M. Heterogenous expression of the related MPB70 and MPB83 proteins distinguish various substrains of Mycobacterium bovis BCG and Mycobacterium tuberculosis H37Rv. Scandinavian journal of immunology. 1996;43(4):374-80. PubMed PMID: 8668915.

20. Schnappinger D, Ehrt S, Voskuil MI, Liu Y, Mangan JA, Monahan IM, et al. Transcriptional Adaptation of Mycobacterium tuberculosis within Macrophages: Insights into the Phagosomal Environment. The Journal of experimental medicine. 2003;198(5):693-704. doi: 10.1084/jem.20030846. PubMed PMID: 12953091; PubMed Central PMCID: PMC2194186.

21. Charlet D, Mostowy S, Alexander D, Sit L, Wiker HG, Behr MA. Reduced expression of antigenic proteins MPB70 and MPB83 in Mycobacterium bovis BCG strains due to a start codon mutation in sigK. Molecular microbiology. 2005;56(5):1302-13. doi: 10.1111/j.1365-2958.2005.04618.x. PubMed PMID: 15882422.

22. Brennan PJ. Structure, function, and biogenesis of the cell wall of Mycobacterium tuberculosis. Tuberculosis. 2003;83(1-3):91-7. PubMed PMID: 12758196.

23. Malaga W, Constant P, Euphrasie D, Cataldi A, Daffe M, Reyrat JM, et al. Deciphering the genetic bases of the structural diversity of phenolic glycolipids in strains of the Mycobacterium tuberculosis complex. The Journal of biological chemistry. 2008;283(22):15177-84. doi: 10.1074/jbc.M710275200. PubMed PMID: 18390543; PubMed Central PMCID: PMC3258887.

24. Tabouret G, Astarie-Dequeker C, Demangel C, Malaga W, Constant P, Ray A, et al. Mycobacterium leprae phenolglycolipid-1 expressed by engineered M. bovis BCG modulates early interaction with human phagocytes. PLoS pathogens. 2010;6(10):e1001159. doi: 10.1371/journal.ppat.1001159. PubMed PMID: 20975946; PubMed Central PMCID: PMC2958813.

25. Barroso S, Castelli R, Baggelaar MP, Geerdink D, ter Horst B, Casas-Arce E, et al. Total synthesis of the triglycosyl phenolic glycolipid PGL-tb1 from Mycobacterium tuberculosis. Angewandte Chemie. 2012;51(47):11774-7. doi: 10.1002/anie.201206221. PubMed PMID: 23086739.

26. Bourke J, Brereton CF, Gordon SV, Lavelle EC, Scanlan EM. The synthesis and biological evaluation of mycobacterial p-hydroxybenzoic acid derivatives (p-HBADs). Organic & biomolecular chemistry. 2014;12(7):1114-23. doi: 10.1039/c3ob42277a. PubMed PMID: 24424275.

ORAL

O-02.1

MYCOBACTERIUM AVIUM SUBSP. PARATUBERCULOSIS FATTY-ACID COMPOSITION MIGHT DETERMINE ITS INTERACTION WITH BOVINE MACROPHAGES AND INFLUENCE THE INTRACELLULAR SURVIVAL OF THE BACILLUS

Marta Alonso-Hearn[1], Naiara Abendaño[1], Amparo Rubirar[2], Rosa Aznar[2], Ramon A. Juste[1,3]

[1]Neiker-Tecnalia, Basque Institute for Agricultural Research and Development, Derio, Bizkaia, Spain. [2]Spanish Type Culture Collection (CECT), Identification and Characterization Service of Microbial Isolates, Paterna, Valencia, Spain. [3]SERIDA, Department of Agriculture of the Regional Government of the Principality of Asturias, Deva, Asturias, Spain

Previously, we demonstrated that Mycobacterium avium subsp. paratuberculosis (MAP) isolates from sheep persisted within bovine macrophages in lower CFUs than cattle, bison, deer and wild boar isolates after 7 days of infection, regardless of genotype (Abendaño et al., 2013). In addition, we showed significant differences in the expression of several cytokines (IL-6, TGF-β1, TNF-α2, IFN-γ, IL-1α), and inhibitors of apoptosis or tissue destruction (BCL2-1, MMP3-1) after the infection of BoMac cells with a bovine or an ovine isolate of MAP. In the current study, we hypothesized that these differences could be induced by variations in the fatty-acid composition of the phophatidyl-1-myo-inositol mannosides (PIMs) of the MAP cell wall that mediate recognition by the macrophage receptors (MR). To test this hypothesis, we investigated the responses of two MAP isolates of bovine (C-type) and ovine (S-type) origin to the bovine and ovine macrophage environment by measuring the fatty acid content of extracellular and intracellular bacteria. For this purpose, macrophages cell lines of bovine (BoMac) and ovine (MOCL-4) origin were infected with the two selected MAP isolates for 4 days at 37 °C. The fatty-acid profiles of the two isolates recovered from infected BOMAC and MOCL-4 cells was determined by gas chromatography (GC) and compared with that of extracellular bacteria and that of bacteria grown in Middlebrook 7H9 medium. Using this approach, we demonstrated that the fatty-acid profiles from extracellular and 7H9-grown bacteria were very similar, and statistically different from that of intracellular bacteria. Analysis of fatty-acid composition from extracellular bacteria enabled the distinction of the two MAP isolates based on the presence of the tuberculostearic acid (18:00 10ME) exclusively in the bovine isolate of MAP. In addition, significant differences were observed in the content of oleic acid and palmitic acid and its derivates between both isolates harvested from the extracellular environment. Our results reinforce the idea that the fatty-acids might impact the spatial conformation of the PIMs mannose caps for recognition by the MR and influence the host immune response and the survival of the bacillus within host cells.

Keywords : MAP-host interaction, mannose macrophage receptor, bacterial lipid metabolism

O-02.2

GENES ESSENTIAL FOR MYCOBACTERIUM AVIUM SUBSP. PARATUBERCULOSIS SURVIVAL AND PERSISTENCE FOLLOWING ORAL INFECTION OF DAIRY CALVES

Jeroen De Buck[1], Amanda Mirto[1], Joyce Wang[2], Marcel Behr[2], Herman W. Barkema[1]

[1]Department of Production Animal Health, University of Calgary, Faculty of Veterinary Medicine, Department of Production Animal Health. 3330 Hospital Drive NW, T2N 4N1, Calgary, Alberta, Canada. [2]McGill University, Faculty of Medicine, Department of Microbiology, McGill University Health Centre, 1001 boul Décarie, Montréal, Quebec H4A 3J1, Canada

Background: A genome-wide understanding of MAP genes required to survive and persist in the native host will offer both fundamental insights into the biology of this organism and candidate targets for translational studies.
Methods: Using a transposon donor phagemid containing a mycomar transposon, a library of MAP Tn mutants was created whereby each mutant has a single gene disruption. This method had been previously validated in an in vitro screen of mycobactin-independent growth (Wang, J. Bact, 2016) and an in vivo screen for persistence in C57Bl/6 mice (Wang, BMC Genomics, 2014). In brief, using the Tn as a primer site, targeted next generation sequencing was used to identify disrupted genes and compare the input pool (the inoculum) with the output pool (MAP clones isolated following experimental calf infection). Genes conditionally essential for persistence in the bovine host were inferred by their relative depletion in the output pool.
Results: Fourteen calves were inoculated at 2 weeks of age with 10^11 mutants or an equivalent inoculum of wild type MAP. Following 2 and 4 months of incubation, intestinal tissues from the ileum and jejunum and their associated lymph nodes were collected from the calves. Depending on the tissue assayed, a range of 50 to 5000 cfu/g were isolated from these tissues and grown on solid media. At the time of submission, the input and output libraries have been generated and submitted for HiSeq-based assignment of Tn insertion sites, with final results expected shortly.
Summary: The identification of genes essential for survival in the calf model will stimulate further research into the pathogenesis of Johne's disease and may contribute to the development of vaccines for the prevention of JD.
A comparison of genes conditionally essential across conditions should illuminate the biochemistry of MAP infection; a comparison of genes conditionally essential across animal models should reveal similarities and differences between hosts in the strategies required by MAP for chronic persistent infection.

Keywords : Essential genes, persistence, virulence, KO mutants, whole genome sequencing, experimental infection, calf model, vaccine

O-02.3

CELL WALL LIPOPEPTIDES OF MYCOBACTERIUM AVIUM: NEW INSIGHTS FROM GENOMICS ANALYSIS

John P. Bannantine[1], Gilles Etienne[2,3], Françoise Laval[2,3], Anne Lemassu[2,3], Mamadou Daffé[2,3], Darrell O. Bayles[1], Christelle Ganneau[4], Maxime Branger[5], Thierry Cochard[5], Sylvie Bay[4], **Franck Biet**[5]

[1]USDA-Agricultural Research Service, National Animal Disease Center, Ames, Iowa, U.S.A. [2]Institut de Pharmacologie et de Biologie Structurale, Université de Toulouse, CNRS, UPS, FranceCNRS, Institut de Pharmacologie et de Biologie Structurale, 31000 Toulouse, France. [3]Université de Toulouse, UPS, IPBS, 31000 Toulouse, France. [4]Institut Pasteur, Unité Chimie des Biomolécules, CNRS UMR 3523, 75724 Paris Cedex 15, France. [5]INRA, UMR1282, Infectiologie et Santé Publique, F-37380 Nouzilly, France

Mycobacteria have a complex cell wall structure that includes many lipids which are often species specific. Besides giving a phenotypic signature these lipids are often involved in infectious processes of pathogenic mycobacteria by interfering with the host immune system.

The biosynthesis pathway of the simplest glycopeptidolipids (GPLs) is relatively well understood and involves more than fifteen genes. Even though M. avium subsp. paratuberculosis (Map) produces a lipopeptide rather than GPL, its genome contains nevertheless a locus highly similar to the GPL biosynthetic pathway of M. avium subsp. avium (Mav). We showed that the module composition of the non-ribosomal protein synthase (Nrp) of Map, the enzyme involved in the synthesis of the peptidyl moiety, is dramatically different from that of other GPL producers such as M. smegmatis (Ms) and Mav.

While Map isolates do not produce GPLs, they do produce lipopeptides without the carbohydrate moiety. However, the picture is not as clear regarding the diversity of lipopeptides produced among two lineages classified as type I/III or S-type (ovine) and type II or C-type (bovine) Map strains that have emerged from the common ancestor, M. avium subsp. hominissuis. The S-type isolates are readily distinguishable from C-type isolates based on genome studies and readily discriminated by genotyping methods. In addition to the genotypic distinctions between S- and C-type strains, phenotypic differences have been documented.

To provide a genomic basis for the synthesis of the diversity of lipopeptides in Map, its recently published genome sequence was explored using in silico methods and completed by biochemical investigations.

Interestingly we discovered a change in the chemical structure of the lipopeptide of the S strains. These findings add new phenotypic evidence that contribute to separate the S type to the C type. Furthermore deciphering the biosynthesis pathway of cell wall lipopeptides should contribute to better understand the determinants of the adaptation of a pathogen to a specific host but also the factors favoring transmission to a new host.

Keywords :
Cell wall, lipopeptide, glycopeptidolipids, non-ribosomal protein synthase

O-02.4

PROTEINS OF MYCOBACTERIUM AVIUM SUBSPECIES PARATUBERCULOSIS THAT BIND AND HYDROLYZE PEPTIDOGLYCAN

John Bannantine[1], Judith R. Stabel[1], Cari K. Lingle[2], Philip R. Adam[3], Kasra X. Ramyar[4], William J. McWhorter[4], William D. Picking[3], Brian V. Geisbrecht[4]

[1]USDA-Agricultural Research Service, National Animal Disease Center, Ames, IA. [2]School of Biological Sciences; University of Missouri-Kansas City, Kansas City, MO. [3]Department of Microbiology & Molecular Genetics; Oklahoma State University, Stillwater, OK. [4]Department of Biochemistry and Molecular Biophysics, Kansas State University, Manhattan, KS, USA

The MAP genome encodes a total of five NlpC/P60 domain-containing proteins, which have been designated MAP_0036, MAP_1203, MAP_1204, MAP_1272c and MAP_1928c in the bovine strain K-10. This is the same number of NlpC/P60 proteins annotated in the M. tuberculosis and Corynebacterium glutamicum genomes. These NlpC/P60 proteins are bacterial peptidoglycan hydrolases that cleave peptide linkages and contribute to cell wall remodeling as well as cell separation during late stages of division. The atomic resolution crystal structures of MAP_1272c and MAP_1204 were determined. These crystal structures, combined with functional assays to measure peptidoglycan cleavage activity, led to the observation that MAP_1272c does not have a functional catalytic core for peptidoglycan hydrolysis. Furthermore, the structure and sequence of MAP_1272c demonstrate that the catalytic residues normally required for hydrolysis are absent, and the protein does not bind peptidoglycan as efficiently as MAP_1204. While the NlpC/P60 catalytic triad is present in MAP_1204, changing the catalytic cysteine-155 residue to a serine significantly diminished catalytic activity, but did not affect binding to peptidoglycan. Collectively, these findings may suggest a broader functional repertoire for NlpC/P60 domain-containing proteins than simply hydrolases.

Keywords :
Peptidoglycan, Proteins, Antigens, Crystal structure

O-02.5

CONSTRUCTION AND CHARACTERIZATION OF A LPRG MUTANT OF MYCOBACTERIUM AVIUM SUBSP PARATUBERCULOSIS K10

Viale Mariana Noelia[1], Park Kun Taek[2], Imperiale Belén[1], Alonso Natalia[1], Colombatti Olivieri María Alejandra[1], Davis William[2], Romano María Isabel[1], **Santangelo María de la Paz**[1]

[1]Instituto de Biotecnología, Instituto Nacional de Tecnología Agropecuaria, Hurlingham, Buenos Aires 1686, Argentina.
[2]Department of Veterinary Microbiology and Pathology, College of Veterinary Medicine, Washington State University, Pullman, WA 99164, USA

The identification and characterization of mycobacterial antigenic proteins is important for the development of new diagnostic tests and vaccines, and for the understanding of the pathogenic mechanisms of mycobacteria and the host´s immune response. The lprG-p55 operon of Mycobacterium tuberculosis and Mycobacterium bovis facilitates the intracellular pathogen multiplication, exacerbating the course of infection. LprG is a lipoprotein that modulates the host's immune response against mycobacteria, while P55 is an efflux pump that provides resistance to several drugs. The knockout mutation of this operon severely reduces the replication of both mycobacterial species during infection in macrophages and mice, and increases susceptibility to toxic compounds. lprG was also described in the Mycobacterium avium complex as an antigen of 22KDa. Our results using Mycobacterium avium subspecies avium as a model, suggested that functional LprG and P55 are necessary for the correct transport of toxic compounds and for the survival of bacteria in in vitro and in vivo models.
In order to gain insight into the function of LprG in Mycobacterium avium subspecies paratuberculosis (MAP), we constructed a mutant using the K10 strain. A phage-mediated allelic exchange technique was used to replace lprG gene with a hygromicin resistant cassette. PCR with the specific lprG primers was performed to confirm the allelic replacement and absence of lprG in those colonies that were hygromicin resistant. The transcription of lprG was evaluated conducting RT-PCR that revealed the absence of transcripts encoding LprG. To confirm that the chromosomal mutation disrupted LprG synthesis, cell lysates from MAP wild type and lprG mutant candidates were analyzed by SDS-PAGE and western blot using anti-LprG polyclonal mice sera. A 22-KDa band corresponding to the expected size for LprG was observed in whole-cell extracts from the wild type strain but it was absent in the mutant strain. These results confirmed the disruption of the lprG gene and the absence of the protein in the mutant strain. In vitro assays are ongoing to test the effect of the lprG gene knockout in the virulence in macrophages and the susceptibility to toxic compounds in MAP.

Keywords :
Virulence, mutants, lprG

O-02.6

NEW WGS DATA OF MAP-S STRAIN JIII-386 FROM GERMANY AND COMPREHENSIVE GENOME ANALYSIS UNVEILED NOVEL CDS, REGULATORY ELEMENTS AND POLYMORPHISM REGIONS

Möbius P.[1], Hölzer M.[2], Felder M.[3], Nordsiek G.[4], Groth M.[3], Köhler H.[1], Reichwald K.[3], Platzer M.[3], Marz M.[2]

[1]*Institute of Molecular Pathogenesis, Friedrich-Loeffler-Institut (Federal Research Institute for Animal Health), Jena, Germany.* [2]*Faculty of Mathematics and Computer Science, Friedrich-Schiller-University Jena, Germany.* [3]*Leibniz Institute on Aging - Fritz-Lipmann-Institute, Jena, Germany.* [4]*Department of Genome Analysis, Helmholtz Centre for Infection Research, Braunschweig, Germany*

Mycobacterium avium subsp. paratuberculosis (MAP) - the etiologic agent of paratuberculosis - affects cattle, sheep and other ruminants worldwide. To decipher phenotypic differences including virulence and host association observed between cattle and sheep strains (MAP-C and MAP-S), data of isolates originating from different geographic regions of the world are required.
MAP sheep strain JIII-386 (MAP-S) from a migrating herd in Germany was subjected to whole-genome shotgun sequencing, de novo assembled, and annotated by BacProt. NcRNAs were annotated by homology search of Rfam families using the GORAP pipeline. Additionally, a new finished sequence of cattle isolate JII-1961 from Germany, published MAP-C strains K10, MAP4 (both from U.S.), MAP-S draft genomes of strain S397 (from U.S.) and strain CLIJ361 (from Australia), as well as M. a. subsp. hominissuis strain MAH104 were used for comparison and assembly improvement of JIII-386. All genomes were annotated by BacProt and results compared with NCBI annotation.
Both approaches detected corresponding protein-coding sequences (CDSs), but also CDSs that were exclusively determined either by NCBI or BacProt. A new Shine-Dalgarno sequence motif was extracted, possibly conserved among Mycobacteria. Novel CDSs and about 80 ncRNAs and Riboswitches were unveiled for MAP; numbers of ASpks, G1, ykkC-III differ among MAP-S and MAP-C. A high sequence conservation of all newly identified ncRNAs was observed. One to two CDSs encoding proteins of PE-PGRS family were identified in MAP as well as in MAH104. Some previously described differences between genomes of MAP-S and -C were partially revised: Four out of ten assumed MAP-S specific large sequence polymorphism regions (LSPSs) are still present in MAP-C strains. Four new LSPSs were identified. Independent of regional origin, results of protein coding gene annotation and single nucleotide variant (SNV) analysis confirm the strong similarity of MAP-C strains, and show higher diversity among MAP-S strains. The calculated phylogenetic trees based on comparison of nucleotide- or amino acid sequences within 790 corresponding CDSs and additionally of corresponding ncRNA sequences give ambiguous results regarding evolution of MAP-C and MAP-S strains.
Comprehensive insights in the MAP genome will help to decipher genes involved in host association and virulence of MAP-S and MAP-C.

Keywords : ncRNA, new Shine-Dalgarno sequence motif, new sheep specific LSPs, SNV/SNP, evolution of MAP-types

O-02.7

PHYLOGENY OF NEW ZEALAND STRAINS OF MYCOBACTERIUM AVIUM PARATUBERCULOSIS BY WHOLE GENOME SEQUENCING

Marian Price-Carter[1], Des Collins[2], Geoff de Lisle[1], Solis Norton[3], Cord Heuer[4], Milan Guatam[4], Jeroen De Buck[5], Christina Ahlstrom[6], Hinrich Vogue[7]

[1]*AgResearch, Palmerston North, New Zealand.* [2]*DM Collins Consulting, Wellington New Zealand.* [3]*Johne's Management Ltd., Dunedin, New Zealand.* [4]*Massey University, Palmerston North, New Zealand.* [5]*University of Calgary, Calgary, Canada.* [6]*Epi-interactive, Wellington, New Zealand.* [7]*LIC, Hamilton, New Zealand*

Background: Recent overseas investigations have demonstrated the far superior ability of whole genome sequencing (WGS) over VNTR/SSR typing to distinguish Mycobacterium avium subsp. paratuberculosis (MAP) subtypes. As a first step in using WGS to assist in the control of Johne's disease in New Zealand (NZ) cattle, deer and sheep farming systems, we aim to use it to define the relationship of important NZ MAP lineages. Specifically, we seek to determine if there is regional clustering of MAP subtypes and if specific subtypes correlate with disease severity.
Methods: We describe the preliminary WGS characterisation of 38 MAP isolates cultured from NZ red deer. Twenty-three isolates were cultured from lymph nodes collected from different regions of NZ, ranging from grossly normal to enlarged and severely lesioned, with an array of different MAP loads (observed acid fast organisms). A further 15 isolates were obtained from faecal samples collected ante-mortem from deer sero-positive for JD and shedding MAP (measured by qPCR) on 4 farms.
Results and Discussion: Preliminary analysis of the single nucleotide polymorphism (SNP)-based phylogeny from these isolates clearly illustrated the superior ability of WGS for distinguishing MAP isolates. Isolates that appeared identical by VNTR/SSR8 typing belonged to 3 distinct clades that differed from one another by 100s of SNPs. Isolates from the same location often shared recent common ancestors, suggesting regional clustering that was undetected by VNTR/SSR8 typing. Although there was no obvious correlation between type and disease severity, a link of this nature may be revealed by alternative methods of data processing and analysis, or with more data. Currently, 60 isolates from below average body condition ewes of Merino and other breeds of NZ sheep, with known and varied histopathology, from farms scattered throughout NZ, and of 39 MAP isolates from dairy cows born on well-established (>10 years old) regionally diverse dairy farms are being assayed by WGS. Analysis of this combined set of NZ MAP isolates will also be described.

Keywords : Whole genome sequencing, genotyping, strain typing, New Zealand, deer, sheep, dairy

O-02.8

The relative frequencies of four major strain types of Mycobacterium avium subsp. paratuberculosis in Canadian dairy herds

Jeroen De Buck[1], Christina Ahlstrom[1], Herman W. Barkema[1]

[1]*Department of Production Animal Health, University of Calgary, Faculty of Veterinary Medicine, Canada*

Background: Whole genome sequencing was previously performed on a representative set of 182 MAP isolates from Canadian dairy herds, identifying nine divergent clades. Four clades were of particular interest as they included either MAP types that have been scarcely reported in North American cattle to-date or represented a significant proportion of isolates. The relatively low numbers of variant sites in the MAP genome are often separated by thousands of base pairs, limiting the use of SNP-based genotyping on a single genomic region.

Aims: The goals of this study were to select SNPs to differentiate four MAP clades of interest by a single PCR, and to determine the relative frequency of MAP clades in all Canadian provinces and to establish a method to inexpensively detect multiple SNPs at once.

Methods: A SNP-PCR assay was developed to facilitate the interrogation of five SNPs located in two distant regions of the genome by linking them together in a single PCR reaction for subsequent Sanger sequencing. This single reaction, high-throughput assay enabled discrimination of 602 MAP isolates from 264 herds, representing all provinces.

Results: Multiple isolates was isolated from 133 of these herds, of which 14 harbored multiple subtypes. The dominant type included 87% of isolates and the Bison type (B) was more distributed than previously identified. A single type III isolate was identified. Type B and the second dominant clade isolates were overrepresented in Québec and Saskatchewan, respectively. Interestingly, the seconc dominant clade was not observed in a diverse set of 141 global MAP isolates suggesting it may be less abundant in other regions.

Summary: The distribution and relative frequency of major Type II subtypes within Canadian dairy herds was assessed, an important step in understanding the clinical relevance and transmission dynamics of MAP in this population and elsewhere.

Applying this SNP-PCR to other regions will identify the relative frequencies of these genotypes outside of Canada. The PCR assay developed in this study can also be modified to target specific clades of interest in other geographical locations.

Keywords :
Whole genome sequencing, genotypes, subtypes, SNP-PCR

O-02.9

Molecular Epidemiology of MAP infection in Chile

Cristobal Verdugo[1], Camilo Tomckoviak[1], Carolina Avilez[1], Miguel Salgado[1]

[1]*Instituto de Medicina Preventiva Veterinaria, Universidad Austral de Chile, Chile*

An epidemiological characteristic of MAP infection is that most infected ruminants never progress toward the clinical stage, remaining sub-clinically infected. This feature of MAP infection, probably involves several factors including host-susceptibility, environment and pathogenic differences between MAP strains. However, little is known about MAP strain diversity among susceptible species. Therefore, the objective was to describe MAP molecular diversity in isolates obtained from different hosts present in Chile and determine possible epidemiological associations between MAP diversity and host species.

A total of 91 independent isolates were sourced from livestock species (cattle, sheep, goat, red deer, & alpaca) and wildlife species (huemul, guanaco, & hares), from different agro-ecological zones of Chile. Isolates were subtyped using a combination of 5 variable number of tandem repeats (VNTR, 292, X3, 25, 47, & 3) and 2 short sequence repeat (SSR, L2 & L8) markers. Subtypes were classified as a composite assignment, combining the results from the VNTR and SSR markers. Rarefaction analysis, Diversity Index (DI), and proportional similarity index (PSI), were used to describe subtype richness and potential associations between subtypes and epidemiological factors.

The combination of VNTR and SSR markers generated 15 subtypes, although 2 subtypes represented 80% of isolates. Those two isolates differentiated only by 1 repetition on marker 292. Dairy cattle was the livestock sector that presented the greatest subtype richness, with 10 subtypes. Nevertheless, the two dominant subtypes represented the 84%. A DI of 0.56 was estimated, which represents a low subtype diversity. Similarly, The PSI comparison indicated a relatively high probability of finding the same MAP subtypes in different host species. These results suggest that MAP in Chile present a low diversity, where infection probably had a common source. This is the first study describing MAP diversity in Chile and the subtypes identified will the basis for further longitudinal studies, addressing the role of MAP strains on clinical disease.

Keywords :
Molecular Epidemiology, Diversity

POSTER

P-02.1
Genetic diversity of Mycobacterium avium sp paratuberculosis isolates from Buenos Aires, Argentina by MIRU-VNTR analysis

Imperiale Belén[1], Moyano Damián[1], Romero Magalí[2], Alvarado Pinedo Fiorella[2], **Santangelo María de la Paz**[1], Travería Gabriel[2], Romano María Isabel[1]

[1]Instituto de Biotecnología, CICVyA INTA, Buenos Aires, Argentina. [2]CEDIVE y Facultad de Ciencias Veterinarias, UNLP, Argentina.

Paratuberculosis (PTB) is an infectious gastroenteritis caused by M. avium subsp. paratuberculosis(Map) that affects mainly ruminants, and it has been postulated a potential zoonotic role in human Cohn's disease. It has a significant impact on the economy. Control of PTB requires a better knowledge of the causative agent and of its epidemiology. Objective. The aim of this study was to determine the diversity of Map strains spreading in Argentina. Materials and Methods. This study included isolated strains obtained from faeces of cattle. Samples were decontaminated using hexadecyl pyridinium chloride and cultured in Herrold´s media supplemented with mycobactin. Mycobacteria were identified by Ziehl-Neelsen stain. IS1311 PCR was performed to identify M. avium complex (MAC) while the IS900 PCR was used to identify Map. Map isolates were classified as type C/S according to IS1311 PCR-REA. Genotyping was performed through the analysis of eight MIRU-VNTR loci (292, X3, 25, 47, 3, 7, 10, 32) previously described by Thibault et al 2007. The analysis was performed using unlabeled and fluorescent labeled primers. The patterns were assigned according with INMV database (http://mac-inmv.tours.inra.fr/) (INRA, Nouzilly). The allelic diversity and the Hunter and Gaston discriminatory index (HGDI) were determined through http://insilico.ehu.es/mini_tools/discriminatory_power/index.php.Results. A total of 61 Map isolates belonged to C type were analyzed. A total of 59 out of 61 (96.7%) isolates could be genotyped and 5 different INMV patterns were found: INMV 1 (42.4%), INMV 2 (28.8%), INMV 11 (20.3%), INMV 8 (5.1%) and INMV 5 (3.4%). The highest discriminatory power (D) was obtained for MIRU 292 and VNTR 7 (0.5 and 0.3 respectively) and the global HGDI for the MIRU-VNTR analysis was 0.7043. Discussion. Most of the isolates included in the study could be genotyped and the frequency of the INMV pattern found in Argentina are in concordance with those previously reported by other authors. INMV 1 and 2 are the main genotypes spreading worldwide. The HGDI was similar to others previous reports (HGDI 0.751, Thibault et al).

Keywords : Mycobacterium avium sp paratuberculosis, genetic diversity, MIRU-VNTR

P-02.2
Paratuberculosis in Latin America: a systematic review

Ana Carolina Silva Faria[1], Isis de Freitas Espeschit[1], David Germano Gonçalves Schwarz[1], Marina Campos Castro Souza[1], Fernando Alberto Paolicchi[2], Ramon Antonio Juste[3], Maria Aparecida Scatamburlo Moreira[1]

[1]Laboratory of Bacterial Diseases, Sector of Preventive Veterinary Medicine and Public Health, Department of Veterinary, Universidade Federal de Viçosa - UFV. PH Rolfs Avenue, University Campus. Viçosa, MG – Brazil. [2]Animal Health Group. INTA, Est Exp Agrop Balcarce – Facultad Cs Agrarias UNMdP. Casilla de Correo 276/ PO Box 276 7620 Balcarce, Argentina. [3]Animal Health Department, NEIKER-Tecnalia Berreaga, 1, 48160 Derio. Bizkaia. Spain.

Latin America is the definition of the group of countries located in America and speakers of Spanish and Portuguese and includes countries in South, Central and North America. Paratuberculosis or Johne_disease is a contagious chronic disease of the gastrointestinal tract that affects domestic and wild ruminants and its etiological agent is Mycobacterium avium subsp. paratuberculosis (MAP). Paratuberculosis is relevant of the socio-economic and public health point of view, justifying the descriptive research of the disease situation and researches in Latin American countries. A survey of 68 articles, published until January 2015, in the Scopus database, Pubmed, Agris, Science Direct, referencing the detection of the agent and the disease in Latin America with no restrictions to the date of the research. The key words used were: paratuberculosis, Mycobacterium avium subsp paratuberculosis, Cattle, milk, wildlife, goat, ovine, dairy and each country_ name written in English. Amongst the 68 studies found in 9 of the 20 Latin America countries, 36.76% (25/68) of the articles were related to Brazil, 17.64% (12/68) in Chile, 11.76% (8/68) in Argentina, 8.82% (6/68) in Colombia, 5.88% (4/68) in Mexico, 2.94% (2/68) in Venezuela, 2.94% (2/68) in Peru and 1.47% (1/68) in Panama and Bolivia. The agent_presence was demonstrated in cattle, goats and sheeps and wild animals such as lagomorphs and guanaco. The techniques most frequently used microbiological culture, PCR and ELISA. The small number of studies may result in overestimated or sub vision of the real situation. Lately many countries are demanding MAPs absence certification for trading of animal products. Given the importance of the agriculture activity in this block attention in terms of research and governments programs should be payed to the disease because is not only compromises the economy, but it can also compromise public health. Acknowledgment. We acknowledge the financial support given by the institutions: Fundação de Amparo à Pesquisa do Estado de Minas Gerais (Fapemig), Coordenação de Aperfeiçoamento de Pessoal de Nível Superior (CAPES) and Conselho Nacional de Desenvolvimento Científico e Tecnológico (CNPq) by financial support. Moreira, M. A. S. is supported by CNPq.

Keywords : PCR, ELISA, dairy, milk

P-02.3

Molecular typing of Mycobacterium avium subsp. paratuberculosis (MAP) isolated from dairy goats in Brazil

Ana Carolina Silva Faria[1], Marina de Castro Campos de Souza[1], Isis de Freitas Espeschit Braga[1], Magna Coroa Lima[1], David Germano Gonçalves Schwarz[1], Ana Paula de Souza Rodrigues[2], Erica Bravo Sales[2], Antonio Augusto Fonseca Junior[2], Maria Aparecida Scatamburlo Moreira[1]

[1]*Laboratory of Bacterial Diseases, Sector of Preventive Veterinary Medicine and Public Health, Department of Veterinary, Universidade Federal de Viçosa - UFV. PH Rolfs Avenue, University Campus. Viçosa, MG - Brazil.* [2]*Laboratory of Molecular Biology - Division of Biosafety. National Agricultural Laboratory – LANAGRO/MG. Ministry of Agriculture, Livestock and Supply. Romulo Joviano Avenue, Pedro Leopoldo, MG. Brazil*

Dairy goat participation in the world and Brazilian agricultural scenario has increased. Minas Gerais state is the third largest producer of goat milk in Brazil. Paratuberculosis is a chronic intestinal disease that mostly affects ruminants and can become a public health problem due to the possible association with Crohn_disease. It is transmitted by ingesting Mycobacterium avium subsp. paratuberculosis (MAP) contaminated food or water. In recent years, with the development of molecular techniques, the disease has been identified with greater precision and more quickly. MAP strains (S-Sheep, Cattle-C and B-Bison) have also been typed. The detection of a circulating MAP strain in a herd is important to develop epidemiological studies and thus establish better strategies to control paratuberculosis. This study aimed to identify and type MAP in dairy goat farms in the Zona da Mata, the main producing region of Minas Gerais state, Brazil. Feces and milk samples of 467 animals were collected, processed, inoculated in Herrold_Egg Yolk Agar (HEYM) and submitted to PCR and REA techniques. Eleven (2.36%) animals were positive for the presence of MAP in four properties and the isolates were characterized as type C strain. It was concluded that MAP is present in dairy goats from properties in Zona da Mata region and that strain type C circulates in the area. This is the first report of MAP typing isolated from dairy goat in Brazil.
AcknowledgmentWe acknowledge the financial support given by the institutions: Fundação de Amparo à Pesquisa do Estado de Minas Gerais (Fapemig), Coordenação de Aperfeiçoamento de Pessoal de Nível Superior (CAPES) and Conselho Nacional de Desenvolvimento Científico e Tecnológico (CNPq) by financial support. Moreira, M. A. S. is supported by CNPq

Keywords :
Enzymatic restriction, Johne disease, bacteria

P-02.4

Optimization of molecular techniques to obtain mutants in Mycobacterium avium subsp. avium and Mycobacterium avium subsp. paratuberculosis

Alonso Natalia[1], Viale Mariana[1], Moyano Damián[1], Colombatti Alejandra[1], Mon Laura[1], Piuri Mariana[2], Jackson Mary[3], Romano María Isabel[1], **Santangelo María de la Paz**[1]

[1]*Instituto de Biotecnologia, INTA. Buenos Aires, Argentina.* [2]*Facultad de Ciencias Exactas y Naturales, UBA. Buenos Aires, Argentina.* [3]*Microbiology, Immunology and Pathology Department. Colorado State University. CO., USA*

Mycobacterial infections are major causes of morbidity and mortality in cattle and are also potential zoonotic agents with implications for human health. Mycobacterium avium subsp.paratuberculosis(MAP) is the causative agent of Paratuberculosis (PTB) in ruminants, a chronic granulomatous disease of the intestines that is transmitted via the fecal oral route, and causes important economic losses in Argentina and worldwide. Despite the implementation of comprehensive animal surveillance programs, it is necessary to emphasize the PTB controls on health systems and also to develop new solutions to control, prevent and eradicate the disease. Therefore, it is crucial to deepen the study of virulence genes involved in the pathogenesis of MAP. It is noteworthy, that obtaining mutants in slow growing mycobacteria is a very difficult process, due to the extremely low rate of recombination and efficiency of transformation. The aim of this study is to optimize the molecular strategies for obtaining mutants in virulence-related genes of M. avium complex (MAC) species. For this purpose, we compared three different strategies for the deletion of the mce4 operon: 1. SacB negative selection technique described by Pelicic and collaborators in 1997 using the pPR27 vector; 2. Specialized transduction using the recombinant phasmid phAE87, described in Bardarov et al., 2002 and Park et al., 2008; 3. Recombineering technique described by van Kessel et al in 2007 that optimizes the efficiency of homologous recombination.Results: in this study we compared three different methods, successfully used in Mycobacteriumtuberculosis complex, for their use in MAC species. So far we have obtained MAA transformed with the plasmid pPR27. However, transformation with the pPR27 plasmid could not be achieved in MAP, probably because this bacterium is difficult to transform. We obtained the recombinant phasmid expressing the upstream and downstream regions of the operon mce4 and the obtention of the mutants in MAA and MAP is in progress. Finally, we are optimizing the conditions for the efficient usage of the recombineering strategy in Map. The pros and cons of the three methods are further discussed.

Keywords :
Virulence, mutants, mce

P-02.5

MIRU-VNTR molecular genotyping of Mycobacterium avium subspecies paratuberculosis isolates from bovine herds across the Republic of Ireland

Marcel de Kruijf[1], Olga N. Lesniak[1], Dermot Yearsley[2], Elvira Ramovic[2], Aidan Coffey[1], Jim O'Mahony[1]

1Cork Institute of Technology, Department of Biological Sciences, Rossa Avenue, Bishopstown, Cork, Republic of Ireland. [2]Central Veterinary Research Laboratory, Department of Agriculture, Food and the Marine, Backweston, Celbridge, Kildare, Republic of Ireland

Introduction: Improvement to dairy herd control strategies for Mycobacterium avium subspecies paratuberculosis (MAP) involves a better understanding of its epidemiology and genetic diversity within MAP isolates among herds. Mycobacterial interspersed repetitive unit and variable number tandem repeat (MIRU-VNTR) has been developed as a simple, rapid and cost efficient molecular typing method to differentiate MAP isolates without the need for sophisticated equipment. The aim of this study was to determine the genomic diversity of MAP isolates across the Republic of Ireland by utilising the molecular MIRU-VNTR typing method.Materials and Methods: A total of 126 MAP isolates across 19 Irish counties were obtained and cultured for 12 to 16 weeks. Genotyping was performed by RT-PCR based on 8 established MIRU-VNTR loci. MAP confirmation was done by IS900 RT-PCR and mycobactin J dependency.Results: Results showed that four INMV groups were observed during this study. INMV 1, INMV 2 and INMV 3 were previously reported, but the unique INMV 116 group has never been reported among herds thus far. The molecular pattern of INMV 116 group showed a difference at the MIRU-VNTR loci 292 and X3. INMV 1 counted 67 MAP isolates (53.2%), INMV 2 counted 57 isolates (45.2%) and INMV 3 and INMV 116 counted 1 isolate each (0.8%). Furthermore, the coexistence of two genotypes within one herd was also observed in two counties. Conclusion: INMV 1, INMV 2 and INMV 3 are observed frequently in Europe and comprised 99.2 % of the total MAP isolates genotyped in this study, indicating that MAP has been a homogeneous population across Ireland with a low level of genetic diversity. INMV 116 has, to the best of our knowledge, never been reported among herds before.

Keywords :
MIRU-VNTR, Mycobacterium avium subspecies paratuberculosis, genotyping

P-02.6

Sub-tipyng of Mycobacterium avium subsp. paratuberculosis strains isolated from wild red deer (Cervus elaphus) in Northern Italy

Garbarino C[1], Ricchi M[1], Leo S[1], Bianchi A[3], Gaffuri A[2], Giacomelli S[3], Savi R[1], Bertoletti I[3], Cammi G[1], Pedrotti L[4], Arrigoni N[1]

1Istituto Zooprofilattico Sperimentale della Lombardia e dell'Emilia Romagna, National Reference Centre for Paratuberculosis, Gariga di Podenzano (PC), Italia. [2]Istituto Zooprofilattico Sperimentale della Lombardia e dell'Emilia Romagna, Diagnostic Section of Bergamo, Italia. [3]Istituto Zooprofilattico Sperimentale della Lombardia e dell'Emilia Romagna, Diagnostic Section of Sondrio, Italia. [4]Stelvio National Park, Italia

Eleven MAP field strains were isolated from red deer (Cervus elaphus), culled in the Stelvio National Park (Italian Alps) within a program for population control, from 2011 to 2014. In this area, wild and domestic ruminants share pasture.
All 11 field isolated strains were classified as Type II, the most common type of MAP in wild ruminants (1). The isolated strains were sub-typed using 8 VNTR loci, showing the same allelic profile INMV1 (http://mac-inmv.tours.inra.fr). This profile is one of the most widespread in Italian domestic ruminants, isolated in Northern Italy cattle (2). Moreover, based on information provided by the INMV database, this profile is also one of the most widespread in the world (28% of the field isolates analysed), both in domestic and wild populations (birds, lagomorphs, rodents, carnivores and ruminants). As previously described in other surveys (3, 4), this evidence suggests the possibility of interspecies transmission between domestic and wild animals and emphasizes the importance of sub-molecular typing for epidemiological studies. Founded by Italian Ministry of Health (E87G12000160001). 1 Carta T, Alvarez J, Perez de la Lastra JM, Gortazar C. 2013. Wildlife and paratuberculosis: a review. Res Vet Sci 94:191-197.2 Ricchi M, Barbieri G, Taddei R, Belletti GL, Carra E, Cammi G, Garbarino CA, Arrigoni N. 2011. Effectiveness of combination of Mini-and Microsatellite loci to sub-type Mycobacterium avium subsp paratuberculosis Italian type C isolates. Bmc Veterinary Research 7.3 Fritsch I, Luyven G, Kohler H, Lutz W, Mobius P. 2012. Suspicion of Mycobacterium avium subsp. paratuberculosis transmission between cattle and wild-living red deer (Cervus elaphus) by multitarget genotyping. Appl Environ Microbiol 78:1132-1139.4 Gerritsmann H, Stalder GL, Spergser J, Hoelzl F, Deutz A, Kuebber-Heiss A, Walzer C, Smith S. 2014. Multiple strain infections and high genotypic diversity among Mycobacterium avium subsp. paratuberculosis field isolates from diseased wild and domestic ruminant species in the eastern Alpine region of Austria. Infect Genet Evol 21:244-251.

Keywords :
Red Deer, Genotyping, VNTR

P-02.7

MOLECULAR CHARACTERIZATION OF A RARE PIGMENTED MYCOBACTERIUM AVIUM SUBSP. PARATUBERCULOSIS TYPE C STRAIN

Célia Leão[1,2], Pedro Barbosa[3], Ana Usié[3], Inês Chaves[3], Ana Amaro[2], Ana Botelho[2], Carlos Pinto[4], João Inácio[5], **Karen Stevenson**[1], António Marcos Ramos[3]

1Moredun Research Institute, Edinburgh, United Kingdom . [2]INIAV - Instituto Nacional de Investigação Agrária e Veterinária, Lisboa, Portugal. [3]CEBAL - Centro de Biotecnologia Agrícola e Agro-alimentar do Alentejo, Beja, Portugal . [4]Serviço de Desenvolvimento Agrário de São Miguel, Ponta Delgada, Açores, Portugal. [5]School of Pharmacy and Biomolecular Sciences, University of Brighton, United Kingdom

Mycobacterium avium subspecies paratuberculosis (Map) is the causative agent of paratuberculosis, affecting a wide range of animals, especially ruminants, with a considerable economic impact. Map strains can be classified in two major groups, Type C (or Type II) and Type S (or Type I/III). Type C are the most frequently isolated strains from a wide variety of hosts including cattle and humans, while Type S strains may have a preference for sheep and goats. Yellow pigmented strains from both Type I and Type III have been isolated from small ruminants from a few restricted geographic regions in the United Kingdom, Spain and Faroe Islands. Here we report the characterization of a rare pigmented Map strain isolated from a goat faecal sample from Azores, Portugal. Growth characteristics and F57 real time PCR indicated that the pigmented isolate was a Map strain suspected to belong to Type C. SNPs characterization analysis confirmed it as Type C strain with SNP profile 3. Whole-genome sequencing of this strain, using the Ion Torrent platform, produced a total of 6,767,537 reads, and 4,612,232 were kept after pre-processing. De novo assemblies were performed with MIRA, yielding 51 contigs. Functional annotations were performed with Prokka for the pigmented strain and two reference strains (type S, S397 and type C, K10). Mappings against K10 strain identified 295 variant sites, while for the S397 mappings 3,748 variants were detected. Variant annotation with SnpEff demonstrated a high number of missense SNPs, particularly when comparing against the S397 strain (1,832). A phylogenetic analysis was performed using other 50 Map genome sequences. Also identified was the presence of LSPA20 and deletion 2 and the absence of LSPA4-II, MAV-14, LSPA18 and GPL cluster in the pigmented strain genome, characteristic for Type C strains. This is the first report of a pigmented Map strain isolated in Portugal. Furthermore, it constitutes the first evidence of the existence of pigmented Type C strains. We hypothesize that this discovery may provide novel clues about the evolutionary traits of Map.

Keywords :
Map, paratuberculosis, pigmented strain, Type C, whole genome sequence analysis

P-02.8

MOLECULAR TYPING OF THE MYCOBACTERIUM AVIUM SUBSPECIES PARATUBERCULOSIS ISOLATES FROM LIVESTOCK FARMS UNDER DIFFERENT MANAGEMENT SYSTEMS IN ARID REGIONS OF WESTERN INDIA (RAJASTHAN): A FIRST REPORT

Mukta Jain[1], K K Chaubey[2], Saurabh Gupta[2], Manju Singh[1,2], S V Singh[2], Manjula Singh[3], Sachin Kumar Verma[1], G K Aseri[1], Manali Datta[4], Neelam Jain[4], Parul Yadav[1], Neeraj Khare[1], Bharti Malhotra[5], Sumit Goyal[5], **J S Sohal**[1]

[1&4]AIMT & AIB, Amity University Rajasthan, Jaipur, Rajasthan, India . [2]Animal Health Division, Central Institute for Research on Goats, Makhdoom, PO – Farah, Dist. Mathura, Uttar Pradesh, India. [3]Indian Council of Medical Research, Ansari Nagar, New Delhi, India. [5]Microbiology Department, SMS Medical College, Jaipur, Rajasthan, India

Mycobacterium avium subspecies paratuberculosis (MAP) is endemic in livestock population of the country. However information on bio-types infecting domestic animals in Rajasthan is non-existent. In the present study our laboratory studied molecular epidemiology of MAP bio-types in the arid Western region (Rajasthan) of country. A total of 62 domestic animals (15- goats, 15- cattle, 15- buffaloes and 17- sheep) were sampled from farm herds and flocks. The cattle and buffalo farms were under intensive, whereas sheep farm was under semi-intensive and the goat farm was maintained on extensive management system (grazing only). Animals were screened by ELISA, fecal PCR, fecal qPCR, blood PCR and culture. Despite the hot arid climate and low density of livestock, none of the farms was free from paratuberculosis infection. Using indigenous ELISA, 51.6% (32) animals exhibited antibodies against MAP. In Fecal qPCR, 25.8% (16) animals were shedding MAP bacilli. MAP bacteremia was observed in 16.1% (10) animals using IS900 PCR. Sensitivity of conventional fecal PCR (8.06%) was comparatively low as compared to qPCR (25.8%). Fecal DNA of animals positive in PCR was subjected to molecular typing by direct IS1311 PCR-REA and it was found that animals were infected with 'Bison type' biotype infecting other states of the country. Maximum positivity was observed in cattle followed by buffaloes, sheep and goats. Study highlights that 3 types of livestock management systems still exist in the state of Rajasthan and intensive management of farms was associated with increased incidence of paratuberculosis. This is the first report on molecular typing of MAP strains infecting domestic livestock population and the study is being extended to cover more farms in the Rajasthan state.

Keywords :
Paratuberculosis, Management Practices, genotypes, Western India

P-02.9

MAC-INMV-PLUS : A WEB APPLICATION DEDICATED TO GENOTYPING OF MYCOBACTERIUM AVIUM SUBSP. PARATUBERCULOSIS STRAINS AND OTHER MAC MEMBERS

Thierry Cochard[1], Maxime Branger[1], Virginie Thibault[2], Philip Supply[3,4,5], Srinand Sreevatsan[6], **Franck Biet1**

[1]INRA, UMR1282, Infectiologie et Santé Publique, INRA centre Val de Loire, F-37380 Nouzilly, France. [2]Anses Laboratoire de Niort, 60 rue de pied de fond, 79024 Niort, France. [3]INSERM U1019, F-59019 Lille Cedex, France. [4]CNRS UMR 8204, F-59019 Lille Cedex, France , Univ Lille Nord de France, F-59019 Lille Cedex, France. [5]Institut Pasteur de Lille, Center for Infection and Immunity of Lille, F-59019 Lille Cedex, France. [6]University of Minnesota, College of Veterinary Medicine, 1971 Commonwealth Avenue, Rm 301E, St. Paul, MN 55108, USA

Background: Genotyping applied to strains of Mycobacterium avium subsp. paratuberculosis (MAP) has become an indispensable tool for epidemiological surveillance of this significant veterinary pathogen. For MAP, multi-locus variable number tandem repeat analysis (MLVA) targeting mycobacterial interspersed repetitive units (MIRUs) and other variable-number tandem repeats (VNTRs) was established using 8 markers. Within the past years this standard, portable, reproducible and discriminatory typing method has been frequently applied alone or in combinations with short-sequence-repeat (SSR) sequencing. As such, a number of genotypes of strains of diverse origins have been identified in laboratories worldwide. With the large diffusion of these genotyping methods standardization between laboratories and knowledge of existing INMV profiles and archiving of newly defined genotypes need to be managed. Objective: To develop a web application called "Mac INMV plus database" including MLVA and SSR. This freely accessible service allows users to compare VNTR and SSR subtype data of their strains analyzed by MLVA and with those of existing reference strains.Results: A freely accessibledatabase was created with genotyping data of strains of MAP and of other M. avium complex (MAC) members. This database gathers all profiles obtained in the laboratory and those published in the literature. The database information include incrementing species and subspecies status, INMV and SSR profiles with corresponding alleles/repeat numbers at each locus, and combined numerical genotypes. The user can consult and query all the existing profiles in order to see if profiles identified in their lab already exist. If profiles are not known, the user can request new attribution numbers online. All information can be exported under EXCEL or PDF format. An extensive documentation regarding the genotyping method, protocols, primer sequences, marker information, and associated isolate specific metadata is available to the users.

Keywords :
Genotyping, MIRU-VNTR, SSR, database

P-02.10

SURVIVAL OF MYCOBACTERIUM AVIUM SUBSP. PARATUBERCULOSIS AFTER EXPOSURE TO COMPOUNDS WITH ANTIMICROBIAL EFFECT MEASURED BY PROPIDIUM MONOAZIDE COUPLED WITH QPCR

Petr Kralik[1]

[1]Veterinary Research Institute, Brno, Czech Republic

Compounds with antimicrobial activity (bacteriocins) secreted by G+ and G- bacteria during milk fermentation are capable to inhibit growth other bacteria. Simple bacteriocins are chemically small of medium peptides or proteins. There is evidence that growth of Mycobacterium avium subsp. paratuberculosis (MAP) is suppressed during the fermentation and storage of fermented milk products. However, published studies considered only the complete effect of the fermentation (presence of bacteriocins, pH, temperature), but there was no information about the effect of bacteriocins themselves. Therefore, the aim of presented work was to assess the effect of bacteriocins secreted by commercially available milk cultures. The effect of bacteriocins was determined by the exposition of three defined MAP isolates to culture media that were previously used for the milk cultures propagation. Supernatants were filtered to make them sterile and used for short and long term exposition to MAP isolates. The viability of MAP was determined by the optimised and validated propidium monoazide treatment followed by the qPCR based on the amplification of fragment F57. Results and outgoing consequences of performed experiment will be presented. The work was supported by the Security Research of Ministry of Interior of the Czech Republic VI20152020044.

Keywords :
qPCR, propidium monoazide, bacteriocins, fermentation

P-02.11

Development and ultra-structural analysis of Mycobacterium avium paratuberculosis biofilm of different clinical isolates

Tarun Kumar Sachan[1], Kundan Kumar Chaubey[1], Saurabh Gupta[1], Krishna Dutta Rawat[1], Naveen Kumar[1], Jagdip Singh Sohal[2], Shivalingappa Yamanappa Mukartal[3], Kuldeep Dhama[4], **Shoor Vir Singh**[1]

[1]Animal Health Division, Central Institute for Research on Goats, Mathura, Uttar Pradesh, India. [2]AIMT & AIB, Amity University Rajasthan, Jaipur, India. [3]Department of Veterinary Microbiology, Veterinary College, KVAFSU, Hebbal, Bangalore. [4]Division of Pathology, Indian Veterinary Research Institute, Izatnagar, Uttar Pradesh, India

Mycobacterium avium subspecies paratuberculosis is a slow growing causative agent of Johne_disease, a chronic insidious enteric infection that affects ruminant population world-wide. Despite ubiquitous nature aeend the fact that MAP has been reported from countries, where ever investigated, there has been very little research efforts to assess its survival in agricultural environment of the country. The aim of this study was to evaluate the ability of MAP to persist as bio-films in-vitro culture media. Thirty different MAP cultures were taken from the repository of Microbiology laboratory that were clinical isolates of goats and human beings and were inoculated into the glass bottle containing MB7H9 media without Tween 80 and containing 10% OADC and mycobactin J. After 3 to 6 months of inoculation a thick water air interface of biofilm was developed. Ruthenium red staining and scanning electron microscopy was done for detailed ultrastructural study and specificity of the bio-film. Scanning electron microscopy (SEM) analysis of bio-films, revealed irregular smooth colony and bacteria encased in a thick matrix of extracellular polymeric substances (EPS). The upper air liquid interface of biofilm was taken for ultra-structural observation of MAP biofilm. Early morphological features of the biofilm were attachment and colonization of bacterial cells then the beginning and deposition of EPS with surface wall. The initial colonizers were, as before, rod in nature, than the more bacteria colonizing the surface were evident. Colonization centred predominantly around the corrugated portions and in the vicinity cracks or grooves were observed. The base of the cusp indentation was heavily colonizedand composed of large cell clusters separated by voids. These bio-films appeared to be more abundant, larger, and thicker in nature, and generally comprised a single morpho-type, mostly rod shaped encased in a thick covering of EPS interspersed with channels. These biofilm appeared to have more abundant extracellular matrix, holding the rods together, interspersed with water channels. Liquid flow occurred in in these water channels, allowing diffusion of nutrients, oxygen, and even antimicrobial agents. The tendency of the bacilli to become arranged together into linear cord-like formations was apparent. Ultrastructural analysis also revealed irregular and smooth crystalline structures, which appeared to be calcifications of biofilm material or the formation of mushroom shaped structures.

Keywords :
Mycobacterium avium subspecies paratuberculosis, biofilm, Ultrastructure, scanning electron microscopy

P-02.12

Induction of pigment production in cattle strains of Mycobacterium paratuberculosis

Thanh Le Nguyen[2], Benjamin M C Swift[1], Catherine E D Rees[2]

[1]School of Veterinary Medicine and Science, University of Nottingham, Sutton Bonington Campus, Loughborough LE12 5RD, UK. [2]School of Biosciences, University of Nottingham, Sutton Bonington Campus, Loughborough LE12 5RD, UK

Mycobacterial pigments, especially carotenoids, have been associated with cellular photoprotection and survival, but the regulation of their production and their physiological role have been largely unstudied. In MAP the study of pigmentation has generally focussed on the taxonomic and identification of different strain types. In this study when the reference cattle strains MAP K10 and ATCC 19851 were cultured on slopes in 30 ml plastic tubes for extended periods of time, intense orange pigmentation was observed. However this was not seen when the same strains was cultured in glass tubes. To rule out contamination, pigmented colonies were subjected to f57 MAP-specific PCR and REA-PCR analysis of the IS1311 sequence. Results indicated that the cells were MAP cattle strains, and these did not produce pigmentation when cultured on fresh media at pH7. However pigmentation could be induced when cells were grown on media adjusted to pH 6.5. The carotenoid biosynthesis pathway has been characterised in Mycobacterium tuberculosis and these were used to interrogate the MAP K10 genome using BLAST. The results identified an intact operon for carotenoid pigment production and therefore these strains do have the genetic potential to produce such pigments. Further work is being performed to determine if transcription of these genes is induced when cells are grown under the conditions found to induce pigmentation of colonies. This observation that MAP C strains are able to produce pigments under the certain environmental conditions suggests that they may play a physiological role in protecting cells against a specific stress condition.

Keywords :
Pigment induction, C strain, acidity

NOTES

NOTES

NOTES

NOTES

NOTES

SESSION 3

Diagnostic & Detection

DIAGNOSTICS AND DETECTION OF MAP: CLINICAL CHALLENGES AND FUTURE ADVANCES.

McKenna SL, Atlantic Veterinary College, University of Prince Edward Island, Charlottetown, Prince Edward Island, Canada

There are many challenges from a clinician's standpoint with respect to the detection of paratuberculosis in production animals. From the very beginning when the first recognition of paratuberculosis or Johne's disease was made in Europe (Johne and Frothingham, 1895), the race was on to find a test for the detection of this disease. The first diagnostic methods developed was the complement fixation test (CFT) (Twort, 1912). It was commonly used on its own and eventually in combination with Johnin skin testing for many years. One of the first choices besides CFT was the agar gel immune diffusion assay (AGID) for detection of antibodies against paratuberculosis (Shermann et al., 1984). Soon after that, enzyme-linked immunosorbent assay (ELISA) gained ground in the serodiagnosis of many disease, including paratuberculosis.

As things have progressed from there, many new testing modalities have come along and with that, different ways of implementing such tests. In fact, it has come to the point that for some clinicians and producers, it has become over-complicated to try and understand how well the tests work, which test they should be using, and how much certainty can they have in the results that are generated. At present, there are some areas of diagnostics and detection that need to be addressed with on-going research. From the standpoint of a clinician that deals with questions on a routine basis, the following are some of the common issues that arise.

Credibility. The quality of being able to give advice that people will believe in or trust. From the standpoint of a clinician, there are a lot of knowledge gaps that need to be addressed to establish more credibility. One of the examples of this is just how early can an animal be infected and how much is the infective dose? The scientific literature as well as abstracts submitted to this ICP meeting demonstrate that we are just beginning to better understand this (Mitchell et al., 2012). Transmission of the pathogen, is it limited to the exposure of calves to adults, or is the potential risk of calf-to-calf transmission real? Corbett et al, will shed some light on this area at this ICP.

Early detection. Commonly the question is raised why certain diagnostic tests cannot be used reliably on younger animals. Are there tests that are being looked at that can aid in early detection of paratuberculosis infection? There are numerous abstracts submitted to this ICP meeting which demonstrate a concerted effort by many laboratories to identify novel antigens for ELISA or IFN-gamma tests. Along with that, secreted or culture filtrate proteins of MAP will be presented at this ICP as potential diagnostic methods.

Rapid detection. Certainly research has brought us a long way from the early days of using solid media for organism detection. Part of this was driven from simple throughput requirements of laboratories. Broth culture techniques certainly facilitated faster detection times. However, even that diagnostic modality may be slowly becoming antiquated. More and more emphasis is being put on faster tests. It started with using capture ELISA on broth media (Shin et al., 2009) to now using high-throughput direct fecal PCR (Plain et al., 2014) and phage assisted PCR looking for paratuberculosis in food products (Botsaris et al., 2010).

Optimal testing. Considering that there are so many tests available and new testing modalities continually being researched, it is very obvious that there really isn't an optimal test for all circumstances. As incredibly frustrating that is for producers and clinicians, it is a reality with paratuberculosis. Comparing one test against another is common, and at this ICP there are numerous presentations and posters that will do just that. However, in some cases it is difficult to trying understand the implications in doing so for various reasons. One of the most common reason for this is that there really is no perfect reference standard. From a data analysis perspective, numerous authors have acknowledged the need for latent class methods for statistical evaluation. Development of new statistical methods has focused on the use of Bayesian methods for estimation of sensitivity and specificity, and the receiver-operating characteristic (ROC) curve. This methodology will be demonstrated by one submission to this ICP, but often it is underutilized.

Positive Outcomes. The aforementioned issues with diagnostics and detection of paratuberculosis does not preclude positive outcomes for a clinician. In fact, it is quite the opposite to that. The positive outlook is that there has been a continual forward momentum in research on MAP and numerous success stories do exist. At times it may almost seem overwhelming the amount of research on-going searching for that elusive optimal test. It is essential that the clinician regard this in a positive light and use it as momentum to continue working with producers to mitigate the impact and spread of MAP.

References.

Johne H A and Frothingham L, 1895. Ein eigenthümlicher Fall von Tuberculose beim Rind. Deutsche Zeitschrift für Tiermedizin und vergleichende Pathologie 21, 438-455.

Botsaris G, Slana I, Liapi M, Dodd C, Economides C, Rees C, Pavlik I. 2010. Rapid detection methods for viable Mycobacterium avium subspecies paratuberculosis in milk and cheese. Int J Food Microbiol. Jul 31;141

Mitchell RM, Medley GF, Collins MT, Schukken YH, 2012. A meta-analyis of the effect of dose and age at exposure on shedding of Mycobacerium avium subspecies paratuberculosis (MAP) in experimentally infected calves and cows. Epidemiol Infect. Feb 140(2):231-46.

Sherman DM, Markham RJ, Bates F, 1984. Agar gel immunodiffusion test for diagnosis of clinical paratuberculosis in cattle. J Am Vet Med Assoc. Jul 15;185(2):179--82.

Shin SJ, Anklam K, Manning EJ, Collins MT. 2009. Rapid mycobacterial liquid culture-screening method for Mycobacterium avium complex based on secreted antigen-capture enzyme-linked immunosorbent assay. Clin Vaccine Immunol. May, 16(5):613-20.

Twort CC, 1912. The agglutination and complement fixation reactions in animals experimentally innoculated with Johne's bacillus, with special reference to the relation of this bacillus to other acid-¬fast bacilli. Zentralblatt für Bakteriologie, Mikrobiologie und Hygiene. 1. Abt. Originale A, Medizinische Mikrobiologie, Infektionskran Kheiten und Parasitologie, 66, 316--318.

ORAL

O-03.1

Using immunological tests for early identification of farms with MAP exposure

Kumudika de Silva[1], Douglas Begg[1], Auriol Purdie[1], Karren Plain[1], Richard Whittington[1]

[1]*School of Life and Environmental Sciences, Faculty of Veterinary Science, The University of Sydney, Australia*

When Johne's disease is suspected or detected in a recently introduced animal, identifying other stock that may have been exposed to MAP quickly will assist in management of the disease on-farm. We have retrospectively analysed data from several experimental trials in sheep and cattle to determine which immune parameters can be reliably used, within a few months of MAP exposure, to differentiate between exposed and non-exposed animals. Sheep and cattle (aged 3-4 months) were orally exposed to either S or C strain MAP respectively. Serum antibodies and MAP-specific IFNγ, IL-10 and lymphocyte proliferation were monitored. For sheep, the data were from four trials of 1-2.5 years duration with a total of 93 exposed and 39 non-exposed animals. For cattle, the data were from two trials of 9 months to 4.5 years duration with a total of 30 exposed and 15 non-exposed animals. Overall, when used soon after exposure, the MAP-specific IFNγ response was the best test for identifying MAP-exposed animals. At 16-22 weeks post exposure, a cut-point of 0.105 for MAP-specific IFNγ SP had 100% specificity and 89.5% sensitivity for detecting MAP-exposed sheep. For cattle, at 15-21 weeks post MAP exposure a cut-point of 0.18 had 71.4% specificity and 80% sensitivity for detecting exposed animals. Thus, for both sheep and cattle, the MAP-specific IFNγ test can be applied within 4-5 months of putative exposure and is effective in identifying young animals (at ~7-9 months of age) that have been exposed to MAP.
This test has the potential to assist with on-farm control strategies and in pre-purchase tests for MAP exposure.

Keywords : Exposure, interferon gamma

O-03.2

Associations between avian purified protein derivative hypersensitivity and MAP antibody ELISA in Irish dairy cows.

Aideen. E. Kennedy[1,2], Noel Byrne[1], Jim O'Mahony[2], Riona. G. Sayers[1]

[1]*Animal & Bioscience Research Department, Animal & Grassland Research and Innovation Centre, Teagasc, Moorepark, Fermoy, Co. Cork, Ireland.* [2]*Department of Biological Sciences, Cork Institute of Technology, Bishopstown, Co. Cork, Ireland*

The single intradermal cervical comparative test (SICCT) utilizes intradermal administration of M. bovis and M. avium subsp. avium purified protein derivatives (bPPD and aPPD) to elicit delayed hypersensitivity responses should exposure to bovine tuberculosis (bTB) have occurred. Comparative skin thickness measurements are taken at injection sites on the neck 72 h post-PPD administration to assess bTB status. Cross-reactivity between SICCT and Mycobacterium avium subspecies paratuberculosis (MAP) antibody ELISA has been reported. The aim of this current study was to identify any relationship between aPPD hypersensitivity reactions and MAP ELISA readings in an Irish dairy herd.
A 139-cow dairy herd was recruited to the study based on a 'clean' historical record of bTB. The herd previously recorded a clinical case of Johne's disease in May 2013. As per Irish statutory requirements, this herd underwent annual SICCT testing. During the 2014 SICCT, the aPPD skin thickness measurement was recorded for each cow. This was subsequently categorized into skin thickness increases of 0mm, 1 to 10mm and >10mm, between the day of test administration to the 72 hour assessment. Blood samples were collected prior to administration of SICCT. Samples were also collected on days 10, 16, 42 and 57 post-SICCT. Serum samples were tested using ID Screen Paratuberculosis Indirect Test. Samples recording S/P ratios of ffl70 S/P were categorised as MAP antibody positive.
Serum ELISA results and continuous aPPD skin measurements were used to create a dataset for statistical analysis. Linear regression was performed to identify associations between 72 hour skin thickness increases and MAP ELISA S/P results at each post-SICCT sampling time-point. Independent variables included breed and parity. Variables recording a significance level of $p<0.5$ were retained.
No bTB positive reactors were identified. At the 72-hour assessment, 50 and 23 animals recorded an increased skin thickness of between 1-10mm and >10mm, respectively, at the site of aPPD administration. The pre-SICCT ELISA analysis recorded no antibody positive cows. On day 10 post-SICCT, 30% (n=42) of the herd tested MAP ELISA positive. Lower prevalence's of MAP seropositivity were recorded thereafter.
Significant associations between increased skin thickness at the aPPD site and increased MAP ELISA S/P ratios were identified at each test time point. This was most notable on day 10 post-SICCT, where each 1mm increase in skin thickness was associated with an increase in S/P ratio of 4.4 ($p< 0.001$)
Conclusion : A significant association between aPPD hypersensitivity and MAP ELISA S/P readings was identified. Further research is required to identify whether utilizing aPPD measurements may identify animals that have previously been exposed to MAP, or alternatively, if MAP antibody ELISA is yielding non-specific responses to environmental M. avium species.

Keywords : SICCT, MAP ELISA, PPD

O-03.3
IMMUNOLOGICAL ANALYSIS OF 35 RECOMBINANT ANTIGENS OF M. AVIUM SUBSP PARATUBERCULOSIS IN MICE AND CATTLE

V. Roupie[1#*], S. Holbert[2#], S. Viart[3], C. Tholoniat[4], B. Leroy[3], D. Cappoen[1], M. Branger[2], F. Jurion[1], M. Govaerts[5], C. Van Den Poel[1], B. Lamoureux[4], J-J Letesson[6], L. Malherbe[4], R. Wattiez[3], F. Biet[2], K. Huygen[1]

[1]*Service Immunology, WIV-ISP-Site Ukkel, Brussels, Belgium. *Present address: Unit "Bacterial Zoonoses of livestock", Operational Direction Bacterial Diseases, Veterinary and Agrochemical Research Centre (CODA-CERVA), Groeselenberg, Brussels, Belgium.* [2]*UMR1282, Infectiologie et Santé Publique, INRA Centre Val de Loire, University François Rabelais of Tours, F-37380 Nouzilly, France.* [3]*Department of Proteomics and Microbiology, University of Mons - UMONS, Mons, Belgium.* [4]*Groupement de Défense Sanitaire Région Centre-Val de Loire, F-36018 Châteauroux, France.* [5]*Department of infectious and parasitic diseases, Faculty of Veterinary Medicine, University of Liège, Belgium.* [6]*URBM, Facultés Universitaires de Namur, Belgium*
 # equal contributors

The goal of this study is to evaluate the immunogenicity and the specificity of 35 novel M. avium subsp paratuberculosis (Map) proteins in mice and bovine models. 23 of these 35 candidates were identified in standard Map ATCC 19698 biofilm cultures on Sauton medium by proteomics and immunoproteomics (Leroy et al, 2007). 6 were identified in in vitro dormancy models based on Sauton cultures submitted to different stress conditions (hypoxia, acidic pH, nutrient starvation or non-toxic NO) (unpublished data). Finally, 6 were identified using an in silico analysis of Map genome (Leroy et al., 2009). The 35 proteins were produced as recombinant histidine-tagged proteins in E. coli.
All proteins were evaluated in BALB/c and C57BL/6 mice intravenously infected with Map ATCC 19698, M. avium subsp. avium (Maa)ATCC 15769 or M. bovis (Mb) AN5. Murine spleen cell IFN-γ production following in vitro stimulation with the proteins and antibody specific proteins were analysed. 14 and 6 of these 35 proteins were respectively evaluated in French and Belgian context. In French context, cattle from one certified Map free herd, one Silirum® vaccinated herd and three Map infected herds were tested in IFN-g release assay and ELISA (IDEXX and ID-vet) whereas in Belgian context, cattle from one Map culture-confirmed herd, five Map free herds and two M. bovis infected herds were tested.
In Map infected mice, 6 of the 35 proteins induced very high IFN-γ production in spleen cell cultures. Although some proteins were recognized more strongly in Map/Maa than in M. bovis infected mice, no real species-specific antigens could be identified. In cattle, among the 14 proteins tested for IFN-γ response in French context, 8 showed significant Spearman correlation with Johnin, with particular high mean S/P ratio response. In Belgium context 3 out of the 6 tested proteins induce a Map specific IFN-γ production.
This study reveals interesting candidates that could be use in the cellular and/or serology diagnosis. One of the most promising candidate is MAP0586c as it protect partially MAP infected BALB/C mice (Roupie et al., 2008) and induces significantly and specific IFN-γ production in cattle.

Keywords : Mice, cattle, IFN-g release assay, Map, M. bovis, 35 Map recombinant antigens

O-03.4
THE EFFECT OF INOCULATION DOSE ON THE PRECISE NUMBERS OF MYCOBACTERIUM AVIUM SUBSP. PARATUBERCULOSIS SHED IN FECES OVER TIME.

Caroline Corbett[1], Jeroen De Buck[1], Herman Barkema[1]

[1]*University of Calgary, Faculty of Veterinary Medicine, Department of Production Animal Health. 3330 Hospital Drive NW, T2N 4N1, Calgary, Alberta, Canada.*

The bacterial load of Mycobacterium avium subsp. paratuberculosis (MAP) in the environment has a large impact on the transmission dynamics on a farm; therefore, accurately quantifying the amount of MAP shed by animals as well as present in the environment is key to our understanding and prevention of new infections. However, quantification of MAP from feces and tissues is notoriously difficult. DNA-based methods, in contrast to culture, do not require the bacteria to multiply for detection, as MAP-specific DNA sequences are targeted directly from the sample. By targeting MAP specific single copy sequences with qPCR and the relatively new droplet digital PCR (ddPCR), accurate quantification of MAP bacteria was achieved. With these novel methods in hand, we aimed to determine the association between inoculation dose and quantity of MAP in fecal shedding. The quantification method was validated in a spiking experiment followed by the direct DNA extraction of the sample and quantification with F57-based qPCR and ddPCR. This method was then used to quantify the amount of MAP bacteria present in fecal samples collected from calves inoculated at 10x107, 5x108, or 10x109 CFU's of MAP in the first 2 weeks, the first month, and the second month after inoculation. Fecal samples were collected daily after inoculation for one week, then monthly and subsequently extracted. The longitudinal pattern and quantity of fecal shedding associated with each inoculated dose will be presented, including the range of variation among calves.
Determining the association between inoculation dose and subsequent shedding is not only important to better conduct experimental trials that can be applied to the field, but the quantification tool can also be used to elucidate areas of increased bacterial load, and individual shedding quantities on farms leading to better understanding and control of transmission.

Keywords : Paratuberculosis, inoculation, dose, shedding, association, quantification, ddPCR, qPCR

O-03.5

Comparative study of six commercial enzyme-linked immunosorbent assays for bovine paratuberculosis diagnostic

B. Gentin[1], L. Foures[2], G. Canteneur[3], J. Anderbourg[4], I. Faes[5], E. Sellal[1]

[1]Biosellal, Lyon, France. [2]GDS55, Verdun, France. [3]GDS57, Metz, France. [4]GDS54, Laxou, France. [5]Segilab – LVD55, Bar-le-Duc, France

Previous comparisons of commercial enzyme-linked immunosorbent assays (ELISA) for bovine paratuberculosis diagnostic have already shown some disagreements in the individual status of tested samples. In a context of paratuberculosis free herd certification or eradication, these variations could have important impacts on the management of the monitored herds. These impacts were evaluated in a comparative study of six commercial ELISAs, by testing 2021 cattles coming from 24 different herds in Région Lorraine (France), equally distributed between meat and dairy herds, divided into 3 categories : 13 herds considered as paratuberculosis negatives (sero-prevalence < 2% - considered as negatives by French regulation) (n=1003), 5 herds considered as low infected (2% < sero-prevalence < 8%) (n=490), and 6 herds considered as highly infected (sero-prevalence > 8%) (n=528). The sero-prevalence of each herd had to be known and stable since at least 2 years, and all the bovines older than 2 were tested. The sero-prevalence in all the herds for each ELISA were calculated and the herds were relocated in corresponding categories. 91% of the samples obtained the same status on the six tested ELISAs. The percent agreement between 2 ELISAs regarding the herds categories fluctuated between 94.8% and 98.7%. 50% to 67% of the herds were qualified in the expected category, depending on the ELISA used. Only 6 herds on the 24 tested were located in the same category by all the kits, but not always in the expected one. 28% of the tests done within the negative herds showed higher sero-prevalences than expected, and 69% of the tests done within the highly infected herds showed lower sero-prevalences than expected. Within the expected negative population, 80% of the positive animals were found positive only on one of the six ELISAs. This percent is divided by 2 within the infected population, which suggest that the concordance between the kits is higher in infected populations. These results show that the choice of the commercial ELISA used for individual testing has a decisive impact on the classification of the herds, and so on the way to manage the paratuberculosis at the herd and national levels..

Keywords :
Paratuberculosis, ELISA, diagnostic, comparative study

O-03.6

Use of liquid culture with IS-900 PCR confirmation on fecal samples to assess exposure of young calves to Mycobacterium avium subsp. paratuberculosis in infected herds

de Marchin Emmanuelle[1], Grégoire Fabien[1], Evrard Julien[1], Houtain Jean-Yves[1]

[1]Regional Association for Animal Identification and Health, Ciney, Belgium

Johne's disease elimination from infected herds takes several years and requires to combine a « test and cull » strategy in adult cows with application of biosecurity measures in young calves in order to prevent new infections.
The acceptance of these measures by breeders is the main difficulty in the field since it often has a significant impact on management habits.
The objective of this study was to investigate in field conditions the use of liquid culture with IS-900 PCR confirmation on fecal samples to assess exposure of young calves to Mycobacterium avium subsp. paratuberculosis (MAP) from infected herds.
Material and methods : In 10 herds involved in the South-Belgian regional Johne's disease control plan and containing at least 10 % of MAP shedders among adult cows, 141 calves were sampled at days 7, 30 and 60 of life. At each sampling, data about the origin of colostrum or milk given, the rearing method (suckling vs non-suckling) applied, as well as the housing-type of the calf, were collected.
Each fecal sample was cultured using para-JEM® liquid culture system for 42 days.
Presence of MAP DNA was assessed using an IS-900 real-time PCR.
Univariate logistic regression was used to assess risk factors associated with calves exposure to MAP.
Results : During this study, MAP could be detected in at least one fecal sample of 24 calves (17%; IC95%: 10.7-23.4). 2 calves were detected positive twice (days 7 and 30).
The rearing method was the only risk factor associated with exposure to MAP in this study: suckling calves were 5.9 (IC95%: 2.3 - 15.2) more at risk to be exposed to MAP during their first two months of life than non-suckling calves.
No significant effect could be observed for the proportion of shedders among adult cows in the herd, neither for the type of colostrum or milk (maternal vs commercial) feeding or the type of housing.
Conclusions : This study showed that liquid culture with IS-900 PCR confirmation is sensitive enough to detect MAP in faeces of young calves in infected herds which opens new perspectives in management of herd infected by MAP.

Keywords :
PCR, para-JEM®, calves, management, exposure

O-03.7

Evaluation of diagnostic sensitivity and specificity of a new ELISA test on serum and milk samples using a Bayesian approach

Alvarez J.[1], **Picasso C.**[1], Patnayak D.[1], Goyal S.[1], Wells S.J.[1]

[1]*Department of Veterinary Population Medicine and Veterinary Diagnostic Laboratory, University of Minnesota, St. Paul, MN, 55108, USA.*

Evaluation of the accuracy of techniques for diagnosis of paratuberculosis due to Mycobacterium avium subsp. paratuberculosis infection is impaired by the lack of an accurate gold standard, particularly in the early stages of infection. In this context, use of Bayesian statistics can be a very helpful tool since it allows the assessment of diagnostic performance without assuming a gold-standard through the introduction of previous knowledge on several parameters in the model. Here, we evaluated the performance of a new diagnostic ELISA test on a large number of milk (n=605) and serum (n=628) samples collected from infected and non-infected cattle herds and part of a well-characterized panel of biological samples. Samples had been analyzed previously with another commercially available ELISA and, in addition, results of fecal culture performed on samples from the same animals were also available, so that three tests were combined in the analysis assuming conditional dependence between the two ELISAs. Scientific literature and expert opinion were combined in order to obtain prior distributions for the sensitivity and specificity of the tests used as comparison and for the prevalence of infection, while non-informative priors were used for the parameters associated with the new test under evaluation. Different models were alternatively implemented (namely a multiple tests-2 populations and a multiple tests-multiple populations model) in order to assess the importance of accounting for variable within-herd population. The new ELISA performed similarly to the ELISA used in the comparison on serum samples (median posterior estimates for Sensitivity and specificity of ≈76% and 97% respectively) but offered a higher sensitivity (≈58%) maintaining a comparable specificity (≈97%) than the ELISA used for comparison. In addition the estimated sensitivity of the culture was lower than expected (median posterior estimates below 50%), perhaps due to the sampled population used in the study. In conclusion, the new ELISA can be a useful test for detection of paratuberculosis particularly in the case of milk samples, in which it offered a superior performance than another ELISA used for comparison.

Keywords :
ELISA, sensitivity, specificity, Bayesian

O-03.8

Comparison of a phage-PCR assay to culture and ELISA for the detection of Mycobacterium avium subs. paratuberculosis in dairy cattle

Gerrard Z.E.[1], Swift B.[3], Davidson R.[2], Hutchings M.[2], Huxley J.[3], and Rees C.[2]

[1]*University of Nottingham, School of Biosciences, Sutton Bonington Campus, Leics, LE12 5RD.* [2]*SRUC, Animal and Veterinary Sciences, Roslin Institute Building, Easter Bush, Midlothian, EH15 9RG.* [3]*University of Nottingham, School of Veterinary Science and Medicine, Sutton Bonington Campus, Leics, LE12 5RD, UK*

Due to the limitations caused by the slow growth of the organism, detection of Mycobacterium avium subsp. paratuberculosis (MAP) and diagnosis of Jonhe's disease is notoriously difficult, with limited sensitivity and specificity being reported. The aim of this cross-sectional study was to determine if enumeration of MAP in milk samples from individual animals using the Phage-PCR method could be used to more readily identify 'high shedders' within a herd. To better understand the results gained, results were compared with standard milk ELISA and milk/faecal culture.
For this study six dairy herds were recruited which varied in size, breed and management practices. Milk samples were collected within a week of the quarterly milk ELISA testing from a random cohort of 100 cows per herd. One sample was sent for MAP culture in an accredited laboratory using ESP Trek, and the other was tested using the Phage-PCR method (Botsaris et al., 2013). Analysis of the results indicates that there is little agreement between the Phage-PCR method and the milk ELISA results, with early results suggesting that the highly positive ELISA cows shed very low levels of mycobacteria into the milk.
A further study was conducted where paired faeces and milk samples were collected from a random 25 cows from each herd. Culture results are still pending, and will be available within 2-3 months. However a small number of milk culture experiments have been completed, and the results indicate that only samples with a high number of mycobacteria detected using the phage-PCR method gave a positive culture result, with samples where less than 100 MAP cells were detected did not give a positive PCR result. This is consistent with the well-characterised problem of decreased sensitivity caused by chemical decontamination of culture samples.
As the culture results become available, we should be able to fully evaluate the value of the phage-PCR method to screen individual animals within a herd to identify high shedders, potentially before they become consistently milk ELISA-positive.

Keywords :
Phage-PCR, rapid detection, shedding rates, culture, ELISA, dairy cattle

O-03.9

TEMPERATURE EFFECT ON SURVIVABILITY OF MYCOBACTERIUM AVIUM SUBSPECIES PARATUBERCULOSIS IN A TROPICAL ENVIRONMENT

Orachun Hayakijkosol[1], Robert Hedlefs[1], Elizabeth Kulpa[1], Wayne Hein[2], Jacqueline Picard[1]

[1]*College of Public Health, Medical and Veterinary sciences, James Cook University, Townsville, Queensland, Australia.* [2]*School of Animal and Veterinary Sciences, The University of Adelaide, Adelaide, South Australia, Australia*

Mycobacterium avium subspecies paratuberculosis (MAP or Mptb) causes Johne's disease in ruminants. Johne's disease is a chronic inflammatory bowel disease which is associated with economic losses around the world. Under temperate conditions, Mptb can survive for more than a year in environment. This study aimed to determine temperature effect to Mptb survivability in tropical conditions using a manual equivalent protocol to the HT-J PCR for DNA. Also, RNA and propridium monoazide (PMA) detections were used to confirm bacterial survivability in environmental samples. Autoclaved soil samples from Dingo, Malanda, Townsville and Charters Towers (Tropical area of Australia) and sterilised faeces mixed with 2 strains of Mptb (Bison and Bovine) were used in this study. First set of samples were placed at four locations in winter season and left in wet and dry conditions for 11 months. A second set of samples were placed at the locations at the beginning of summer. Each month, 2 samples of each Mptb strain and 2 samples of each condition (8 samples) were collected for DNA, RNA and PMA tests. Hourly record of temperature and humidity in each location from data loggers was downloaded each month. Temperature over 30 and 40 degree Celsius data and humidity under 50 rH and 40 rH were collected and analysed in hour per month and how many days per month. Temperature and humidity results were compared to number of bacteria from DNA, RNA and PMA results. Dingo and Townsville temperature and humidity data had similar patterns while Malanda and Charters Towers data showed different patterns. After 11 months of study, results from DNA, RNA and PMA test determined that no Mptb from Malanda and Townsville set 2 samples were detected while Set 1 samples from Dingo and Charters Towers showed high percentages of positive results. However, PCR results showed high survivability of Mptb in the dry condition compared to the wet condition. This study suggests that temperature over 30 degree Celsius and humidity under 50% may have significant effect on survivability of Mptb in the tropical environment. Mptb cannot survive during summer season while some Mptb can live longer in winter and dry condition.

Keywords :
Temperature, Humidity, Detection, Mycobacterium avium subspecies paratuberculosis, Environmental samples

O-03.10

INFORMATIVE VALUE OF ENVIRONMENTAL AND POOLED SAMPLES TO EVALUATE THE WITHIN-HERD PREVALENCE OF MYCOBACTERIUM AVIUM SUBSPECIES PARATUBERCULOSIS IN GOATS FLOCKS

Esthel Frederic[1], Valentine Grebert[2], Raphael Guatteo[3], Serge Nouzieres[4]

[1]*GDS18, Bourges, France.* [2]*Clinique Vétérinaire de Bracieux, Bracieux, France.* [3]*UMR Oniris-INRA 1300 BioEpAR, Nantes, France.* [4]*GDS41, Blois, France*

Introduction: Due to both mean size of goat flocks and the value of a single goat, individual testing is considered as an expansive method to evaluate within herd prevalence of paratuberculosis. The aim of this study was to investigate the informative value of PCR applied in environmental and pooled samples to estimate the within-herd seroprevalence and prevalence of Map shedding among lactating goats.
Materials and methods: The study was realized on 697 goats in 19 herds. Milk line filter, bulk tank milk and dusts collected on the milking parlour with dry swab socks were tested using liquid media culture para-JEM® followed by PCR. The results of these environmental and pooled samples were compared with within-herd serological and shedding prevalence. Environmental and pooled samples were considered in a first step alone and then in combination to optimize sensitivity and specificity using ROC curve analysis.
Results: Within-herd serological prevalence ranged from 0 to 0.39; within-herd shedding prevalence ranged from 0 to 0.20. At least one PCR positive result either on the milk line filter or on the bulk tank milk sample allowed to identify herds that had more than 10 % of shedding goats and more than 10 % seropositive goats with a sensitivity of 0,86 and a specificity of 0,75. Two PCR negative results for both the bulk tank milk sample and the dry swab socks allowed to identify herds with less than 5 % shedding goats and less than 5 % seropositive goats with a sensitivity of 0,80 and a specificity of 0,93.
Conclusions: These results confirm that PCR applied on environmental samples could be a relevant alternative to estimate at lower cost and at a large scale the infection status of goats herds towards Map infection.

Keywords :
Goat, detection, environmental, pool, milk, socks, prevalence

O-03.11

Evaluation of the use of a polymerase chain reaction assay on overgrown environmental samples cultured for Mycobacterium avium ssp. paratuberculosis

Juan C. Arango-Sabogal[1], Olivia Labrecque[2], Julie Paré[3], Julie-Hélène Fairbrother[2], Jean-Philippe Roy[1], Vincent Wellemans[1], Gilles Fecteau[1]

[1]Department of Clinical Sciences, Faculty of Veterinary Medicine, Université de Montréal, Saint-Hyacinthe, Québec, Canada. [2]Laboratory of Epidemiological Animal Surveillance of Québec, Ministry of Agriculture, Fisheries and Food of Québec, Saint-Hyacinthe, Québec, Canada. [3]Canadian Food Inspection Agency, Saint-Hyacinthe, Québec, Canada

Culture of Mycobacterium avium ssp. paratuberculosis (MAP) is the definitive antemortem diagnostic method for paratuberculosis. Microbial overgrowth is a challenge for MAP culture, however, as it complicates, delays, and increases cost of the process. Additionally, herd status determination is impeded when non-interpretable (NI) results are obtained. The performance of the polymerase chain reaction (PCR) is comparable to fecal culture, thus it may be a complementary diagnostic tool to classify NI samples. This study aimed to determine if MAP DNA can be identified by PCR performed on NI environmental samples and to evaluate the performance of PCR before and after the culture of these samples in liquid media. A total of 154 environmental samples (62 NI, 62 negative, and 30 positive) were analyzed by PCR before being incubated in an automated system. Growth was confirmed by acid fast bacilli stain and then the same PCR method was again applied on incubated samples, regardless of culture and stain results. Change in MAP DNA after incubation was assessed by converting the PCR quantitative cycle values (Cq) into fold change using the 2-ΔCq method (ΔCq = Cq after culture – Cq before culture). A total of 1.6% (SE=1.6) of the NI environmental samples had detectable MAP DNA. The PCR had a significantly better performance when applied after the culture than before the culture (P=0.004). After the culture, a 66 fold change (SE=17.1) in MAP DNA was observed on average. Performing a PCR on NI samples improves MAP culturing. The PCR method used in this study is a reliable and consistent method to classify NI environmental samples.

Keywords :
Overgrown environmental cultures, Mycobacterium avium ssp. paratuberculosis

O-03.12

A novel ResoLight-based real-time PCR assay with pooled faecal samples for the screening of Johne's disease in cattle

Satoko Kawaji[1], Reiko Nagata[1], Yasutaka Minegishi[2], Yumi Saruyama[3], Akiko Mita[4], Yasuyuki Mori[5]

[1]National Institute of Animal Health, [2]Nippon Gene Co., Ltd. [3]Tochigi prefectural Central Livestock Hygiene Service Center. [4]National Livestock Breeding Center, [5]Zen-noh Institute of Animal Health, Japan

A direct faecal real-time quantitative PCR (QPCR) assay enables rapid and sensitive detection of cattle shedding Mycobacterium avium subsp. paratuberculosis (MAP). For the screening of whole herds, PCR-based tests of individual animals are labour-intensive and more costly than serology, which is mainly used for this purpose even though it lacks sensitivity. In this study, we developed a novel ResoLight-based real-time PCR (RL-PCR) assay combined with an improved DNA extraction for the detection of MAP in pooled faecal samples. The RL-PCR assay included synthetic DNA as an internal amplification control (IC); it is amplified using the same primer pair as the target molecule IS900. The result was interpreted by analysing melting temperature (Tm) peaks; Tm peaks for the target indicated presence of MAP while the Tm peak for IC indicated a reaction without inhibition.

Faecal suspensions were prepared individually and pooled. DNA extraction from pooled samples was performed using an improved version of Johne-Spin (FASMAC). No inhibition of PCR amplification was observed with up to 15 faecal samples in a pool, and with a pool size of 20 slight inhibition was indicated by a reduction in the height of the Tm peak for IC. The detection limit of RL-PCR (pool size=10) was ten MAP organisms per gram of faeces (equivalent to 0.5 MAP per extraction), which was comparable to the limit of QPCR for individual faeces.

The effect of pooling on the sensitivity of the assay was evaluated using 64 QPCR positive bovine faeces. At a pool size of 10 (1 infected plus 9 uninfected), 95.1% of the positive faeces containing ffl the diagnostic cut-off value of 0.001 pg/well of MAP DNA were detected in RL-PCR while the positive rate was 65.2% in the faeces containing less than 0.001 pg/well. Three infected herds were screened by individual antibody-ELISA and the RL-PCR assay using pooled faeces, and the results were compared to the individual QPCR test. In RL-PCR, all the pools which included animals shedding more than 0.001 pg/well of MAP DNA were positive while only 25% were detected by ELISA.

The RL-PCR assay demonstrated high sensitivity for screening of Johne's disease.

Keywords :
Pooled faeces, ResoLight, Real-time PCR, Internal control, Screening test

O-03.13

Detection of Mycobacterium avium subspecies paratuberculosis (MAP) in powdered infant formula

Kamal Raj Acharya[1], Plain K. M.[1], Whittington R. J.[1], Dhand N. K[1]

[1]*Farm Animal & Veterinary Public Health, Faculty of Veterinary Science, School of Life and Environmental Sciences, The University of Sydney, Sydney, New South Wales 2570, Australia*

Concerns have been raised regarding the possible role of MAP in Crohn's disease in humans. Milk derived from animals in the clinical and subclinical stages of paratuberculosis could be contaminated with MAP. Powdered infant formula (PIF), when produced from MAP contaminated milk, could contribute towards early human exposure to MAP or MAP components. Effective tests may be required to ensure PIF are free from MAP. The objective of this study was to develop qPCR and culture methods for the detection of MAP in PIF, and test PIF for the presence of MAP.
A total of 122 PIF samples available in Australia were purchased and tested. The samples comprised 72 brands produced by 12 manufacturers from 9 countries.
PIF samples were reconstituted in ultra-pure water and the casein pellet was separated by centrifugation. An optimised DNA extraction method involving chemical and mechanical lysis followed by a magnetic bead based technique was used. A qPCR test was then performed based on the detection of the IS900 molecular target of MAP. The limit of detection of the test was determined using milk powder spiked with MAP ranging from 100 to 108 MAP per 1.5 grams, with ten replicates each of the concentration tested. Culture of PIF samples in liquid media was also undertaken after the identification of the optimum decontamination protocol among four protocols, evaluated using the most probable number (MPN) technique. After decontamination, cultures were incubated for up to 20 weeks. Confirmation of the presence of MAP at 12 and 20 weeks is currently being undertaken.
Out of 122 samples, 2 were positive on qPCR. The method developed could effectively test PIF for the presence of MAP DNA, with a limit of detection of less than 5 MAP organisms per 1.5g milk powder. This method, applied to dried milk, could facilitate backward tracing of an infected farm, thereby aiding in JD control.

Keywords :
Powdered milk, qPCR, paratuberculosis

O-03.14

Circulating miRNA in bovine serum and their potential use as novel bio markers of early Mycobacterium avium subsp paratuberculosis infection

Ronan G. Shaughnessy*[1], Damien Farrell*[1], Karel Riepema[6], Louise Britton[1], David E. MacHugh[2,5], Bryan Markey[1], Douwe Bakker[6], Stephen V. Gordon[1,3,4,5]

[1]*School of Veterinary Medicine,* [2]*School of Agriculture and Food Science,* [3]*School of Medicine,* [4]*School of Biomolecular and Biomedical Science,* [5]*Conway Institute, University College Dublin, Dublin, Ireland.* [6]*Department of Bacteriology and TSEs, Central Veterinary Institute of Wageningen University, Edelhertweg 15, 8200 AB Lelystad, The Netherlands.*
* *Equal author contribution*

Johne's disease (JD) is a chronic enteritis of ruminants caused by Mycobacterium avium subspecies paratuberculosis (MAP). New disease control tools are needed as current strategies are hampered by the lack of sensitive and specific diagnostic modalities. Circulating microRNAs (miRNAs) have been shown to have potential as novel biomarkers for various human diseases, but their potential application in the veterinary sphere has not been explored. The aim of this study was to use RNA-sequencing to assess the potential of miRNAs as biomarkers for JD disease progression.
Sera from a six month experimental infection conducted at UCD consisting of MAP-challenged calves (n=6) and age-matched controls (n=6) was used. We also analysed sera that had been stored at -20°C for over a decade from a four year MAP infection model performed by the Central Veterinary Institute (CVI) of Wageningen University. Cattle defined as seropositive for anti-MAP antibodies (n=5) were compared against sero-negative cattle (n=7). Comparison of UCD and CVI samples also enabled us to assess the stability of miRNA profiles in biobanked sera.
From the UCD samples, we identified 100-200 known miRNAs with multiple isomiRs and 30 novel miRNAs. Surprisingly, the small RNA profile was highly similar to that of the biobanked CVI samples. No significant differential miRNA expression was detected between MAP-infected and their age-matched controls in either experiment. In contrast, comparing pre-infection sera of young calves to the 6 month interval of each experiment revealed miR-29a, miR-92b and miR-205 increases (2-fold) that may be due to blood-cell population changes during calf maturation ($P<0.001$).
Our study has demonstrated the stability of miRNA in serum, identified a range of novel miRNA in bovine serum, and shown the utility of small RNA sequencing approaches to explore the potential of miRNA as novel biomarkers for infectious disease in cattle.

Keywords :
Bovine infection, miRNA stability, small RNA-sequencing, biomarker

O-03.15
Diagnostic potential of a peptide-mediated magnetic separation (PMS)-phage assay applied to milk relative to faecal- and blood-based tests

Irene Grant[1], Lorna O'Brien[1]

[1]*Institute for Global Food Security, Queen's University Belfast, Northern IrelandSam Strain, Animal Health and Welfare Northern Ireland, Armagh, Northern Ireland*

The PMS-phage assay is a novel rapid test to detect viable Mycobacterium avium subsp. paratuberculosis (MAP) in milk. MyOne™ tosylactivated Dynabeads® coated with MAP–specific biotinylated aMp3 and aMptD peptide binders are added to the resuspended pellet obtained after centrifugation of 50 ml milk. During the 30 min 'capture' in a Dynal BeadRetriever MAP cells bind to the surface of the beads. The residual milk components are removed with two washes before beads and adhering MAP cells are resuspended in Middlebrook 7H9 broth in readiness for the phage assay. The phage assay involves a mycobacteriophage which will only amplify within viable MAP cells causing lysis thereby produce zones of clearing (plaques) in an agar lawn of rapid-growing M. smegmatis. The presence of MAP DNA in plaques, confirmed by IS900PCR, indicates a MAP positive milk sample.
The diagnostic potential of the PMS-phage assay was assessed by testing milk samples collected from individual cattle categorised as either 'MAP-infected' or 'Non-infected' on the basis of prior faecal culture and serum- or milk-ELISA testing. A total of 170 x 50 ml milk samples were tested - 131 from 'non-infected' cows and 39 from 'MAP-infected' cows. Contingency tables (2x2) were constructed comparing the PMS-phage assay results to each of the other diagnostic tests to determine kappa agreement. The diagnostic specificity (Sp), sensitivity (Se), positive predictive value (PPV), and negative predictive value (NPV) of the PMS-phage assay were calculated by constructing a contingency table (2x2) comparing PMS-phage assay result and designated infection status of the animal.
PMS-phage assay results indicated the presence of viable MAP in 19 (11.2%) of the 170 milk samples; 12 were from animals designated 'MAP-infected' and 7 were from animals designated 'non-infected'. There was 'poor' agreement between PMS-phage assay and serum-ELISA results (Kappa 0.047, 95% CI: -0.124 to 0.217) and 'fair' agreement between PMS-phage assay and faecal culture results (Kappa 0.350, 95% CI: 0.163-0.536). Diagnostic Sp, Se, PPV and NPV of the PMS-phage assay applied to milk were 0.9470, 0.3158, 0.6316 and 0.8278, respectively. The PMS-phage assay was shown to have good specificity however sensitivity would need to be improved for diagnostic applications.

Keywords :
PMS-phage assay, milk testing, diagnostic potential

O-03.16
Specific and sensitive detection of viable MAP within 6 h using bacteriophage

Ben Swift[1], Le Van Hung Vuong [1], Catherine Rees[1]

[1]*University of Nottingham, School of Biosciences, UK*

The bacteriophage amplification assay has been shown to be able to rapidly detect viable Mycobacterium avium subsp. paratuberculosis (MAP) in a range of matrices such as milk, cheese and - most recently - in clinical blood samples. However the assay in its current format is laborious and time consuming, meaning it can be difficult to test large volumes of samples in a high-throughput capacity. Using novel technology invented at the University of Nottingham, the original phage assay has been condensed into a one tube format capable of sensitively and specifically detecting MAP from blood and milk within 6 h. The assay is based on using the same phage (D29) as in the original assay but using it as a mycobacterial lysis reagent that will only release mycobacterial DNA from viable cells. The DNA can then be separated from the cell debris and detected using any appropriate downstream molecular detection method. Using this new phage assay we were able to routinely detect two MAP cells per ml of blood and milk in experimentally spiked samples. We have then used the assay to detect MAP in the blood of naturally and experimentally infected cattle. The One Day phage assay is a major breakthrough with regards to detect viable MAP and has the potential to revolutionize detection of viable cells leading to the better control and understanding of both the organism and of Johne's disease.

Keywords :
Bacteriophage, detection, viable, blood, milk

O-03.17

RECOMBINANT SECRETORY PROTEINS BASED DIAGNOSIS FOR DETECTION OF DIFFERENTIAL IMMUNE RESPONSE TO MYCOBACTERIUM AVIUM SUBSPECIES PARATUBERCULOSIS INFECTION AND VACCINATION IN DOMESTIC LIVESTOCK

Kundan Kumar Chaubey[1], Saurabh Gupta[1], Jagdip Singh Sohal[2], Sujata Jayaraman[2], Bjorn John Stephan[1], Manju Singh[1], Mukta Jain[2], Naveen Kumar[3], **Shoor Vir Singh**[1]

[1]Microbiology laboratory, Animal Health Division, Central Institute for Research on Goats, Makhdoom, PO-Farah, Mathura, 281122, India. [2]Amity Institute of Microbial Technology, Amity University Rajasthan, Kant Kalwar, NH 11C Delhi-Jaipur Highway, Jaipur, 303 002, Rajasthan. [3]Veterinary Type Culture Collection, National Research Centre on Equines, Hisar, Haryana-125001, India

Paratuberculosis (pTB) is chronic non-treatable granulomatous enteritis caused by Mycobacterium avium subsp. paratuberculosis (MAP). Control and eradication of pTB using vaccine is difficult in Indian conditions due to lack of efficient indigenous diagnostic tests to differentiate between infected and vaccinated animals (DIVA). Serological tests like ELISA easily detect MAP antibodies induced during infection. However it cannot differentiate whether antibodies generated are due to infection or vaccination. Therefore, study proposed to develop a 'new diagnostic test' based on recombinant secretory proteins which can open a fresh approach for early detection of the disease and which can differentiate infected and vaccinated animals. Secretory proteins are the major tools that bacteria use to interact with their extracellular environment, are highly immunogenic or immuno-dominant, encounter sensitized immune cells. Comparing the secretory proteins with other cellular proteins showed greater sero-reactivity from MAP infected animals. A range of new immunologically important MAP secreted antigens have been identified using new technologies like 2D-gel electrophoresis, chromatography, mass spectrometry and Peptide mass fingerprints. Of the total secretory proteins, at least six (MAP1693c, MAP 2168c, Mod D, Antigen 85c, Pep AN and Pep AC) have been found to react serologically. These six secretory proteins of MAP were successfully cloned, expressed and purified as a His-tag fusion protein in pET28a(+) and pET22b(+) plasmid backbone exhibiting approx 20 kDA, 23 kDA, 40 kDA, 45 kDA, 19 kDA and 24 kDA molecular masses, respectively, was evident from SDS-PAGE and showed high sero-reactivity with positive control sera on immune-blotting. These resulting recombinant MAP secretory proteins will be used as tapping antigen in subunit / cocktail ELISA to investigate differential immune response to infection and vaccination against MAP in domestic livestock. These antigens are potentially more functional that may serve as a DIVA based future diagnostic for monitoring and control of pTB progression. Work using these proteins is in progress.

Keywords :
Mycobacterium avium subspecies paratuberculosis, pTB, DIVA, secretory proteins, sero-reactivity, immune response

POSTER

P-03.1
Incubation time for Mycobacterium avium subsp. paratuberculosis in liquid medium (M7H9C) from bovine clinical samples

Romero, Magali Andrea[1]; Alvarado Pinedo, María Fiorella[1]; Moyano, Roberto Damian[2]; Santangelo, María De La Paz[2]; **Travería, Gabriel Eduardo**[1]

[1]*Centro de Diagnóstico e Investigaciones Veterinarias (CEDIVE), Facultad de Ciencias Veterinarias. Universidad Nacional de La Plata, Buenos Aires, Argentina.* [2]*Instituto de Biotecnología, CICVyA, INTA, Buenos Aires, Argentina*

Paratuberculosis is a chronic devastating ileocolitis of ruminant caused by Mycobacteriumavium subsp paratuberculosis (Map), with unthrifty and diarrhea as the main clinical signs in cattle. Microbiological culture is the gold standard test for paratuberculosis, liquid medium is recognized as having better performance than solid Herrold's egg yolk medium, here we present the incubation time to detect Map in clinical samples coming from naturally infected herds. The liquid culture medium was prepared as previously described except for the addition of a small amount of agar, the presence of acid fast bacilli (AFB) with Ziehl Neelsen method was used as growth indicator at weekly interval. Earlier cultures were obtained from the mucosa of necropsied animals with advanced clinical signs, essentially gut mucosa were positive for AFB 14 days after incubation, followed by mesenteric lymph node and fecal samples. Prolonged incubation times were needed in samples coming from young asymptomatic animals aged 14 months in this group AFB were observed after a period of 1,5 months, probably representing a reduced number of Map present in clinical samples.

Keywords :
Paratuberculosis, culture, liquid media

P-03.2
Detection and quantification of Mycobacterium paratuberculosis by qPCR in ovine faeces: methodology and improvement of fluorescence curves interpretation

Mathevon Y.[1,2], Foucras G.[1,2], Corbiere F.[1,2]

[1]*INRA, ENVT, UMR 1225, IHAP, 31076 Toulouse, France.* [2]*University of Toulouse, INP, ENVT, 23 chemin des Capelles, 31076 Toulouse, France*

Since several years, quantitative polymerase chain reaction (qPCR) has emerged for detection and quantification of Mycobacterium avium subsp. paratuberculosis (MAP). Compared to culture-based technics, qPCR is quicker, more sensitive and allow precise quantification of the bacterial charge in a large variety of biological samples, including faeces. Interpretation of raw qPCR data most of the time relies on unknown algorithm or manual determination of the cycle threshold (CT) and reaction efficiency, when provided, is computed at the qPCR plate level. In this study, we analysed and compared the results for qPCR runs using commercial method implemented in the Lightcycler480's software (LC480, Roche) and using a dedicated package (qpcR) developed for R.Individual faeces from 3900 ewes were sampled in 30 flocks known to be infected or free from paratuberculosis. MAP DNA was extracted from 10 grams of faeces with addition of an extraction control fluorophore using the Adiapure protocol (Adiavet, Biomérieux) and amplification was then performed using Adiavet paratb real time kit for 45 cycles. At the end of the run, a colour compensation step was made to separate the specific IS900's target fluorescence (FAM) from that of extraction control (VIC). Quantification was performed using an external standard with 4 dilution-points run in triplicate. Only samples with an extraction control value under 40 were retained for further analysis. Fluorescent data were analysed using both the LC480 software second derivative maximum algorithm and qpcR package in R, which allows a precise fitting of each sample fluorescence curve via a model selection process. This package also offers new CT definitions such as Cyo and the possibility to go further than the forty cycles limit imposed by the LC480. Using known CT, we defined the analytical sensibility of our amplification process and found that the biological detection limit was around 43.1 cycles. The distribution of qPCR positive animals was clearly bimodal, one peak corresponding to heavy shedders (mean CT value around 28 cycles) the other, around 40 cycles, presenting light shedders or animals with passive transport of MAP. This analytical procedure meets MIQE requirements and permits interpretation of qPCR results in a comprehensive way.

Keywords :
Ovine, qPCR, methodology

P-03.3

Development of the two DNA extraction methods for Mycobacterium avium subsp. paratuberculosis (MAP) from bovine and buffalo bulk milk and colostrum

Fabrizio Gamberale[1], Gabriele Pietrella[1], Valeria Antognetti[1], Paola Scaramella[1], Marcello Sala[1], Antonella Cersini[1]

[1]*Istituto Zooprofilattico Sperimentale del Lazio e Toscana "M. Aleandri", Rome, Italy*

To reduce the Mycobacterium avium paratuberculosis (MAP) in buffalo and pasteurised bovine milk, six different DNA extraction methods have been under examination: the commercial QIAamp® DNA Blood Mini kit (Qiagen); a combination of the commercial QIAamp® DNA Blood Mini kit (Qiagen) with the magnet beads specifically designed for MAP; the commercial NucleoSpin® Food kit (Macherey-Nagel); a combination of the commercial NucleoSpin® Food kit (Macherey-Nagel) with the magnet beads; Chelex® 100; a combination of Chelex® 100 with the magnet beads. Out of these six methods, only two have been selected: the NucleoSpin® Food kit (Macherey-Nagel) and the NucleoSpin® Food kit (Macherey-Nagel) combined with the magnet beads, according to the DNA extraction efficiency, to the DNA purity and DNA suitability for amplification. These two extraction methods have been tested on pasteurised milk cattle. Carrying out the combination of the NucleoSpin® Food kit with the magnet beads, the DNA concentration resulted to be 49,43 ng/µl, while using the NucleoSpin® Food kit, it was 36,7 ng/µl whreas the DNA purity measured by spectrophotometer was 1,80 (by calculating the ratio of absorbance at 260-280 nm) for both protocols. In the NucleoSpin® Food kit applied on pasteurised bovine milk, the limit of detection (LOD) was included between 19,65 x 106 and 18,97 copies of IS900 equivalent, respectively, at 1,31 x 106 UFC and 1,26 UFC (considering that there is on average 15 copies of IS900 per single MAP cell). The NucleoSpin® Food kit combined with the magnet beads specifically designed for MAP, applied on pasteurised bovine milk, the limit of detection (LOD) was included between 16,75 x 106 and 19,70 copies of IS900 equivalent, respectively, at 1,17 x 106 UFC e 1,31 UFC. The same analytical validation methods have been applied for bovine and buffalo milk. The same extraction methods will be applied to both bovine and buffalo colostrum.References: Evaluation of different methods for DNA extraction from milk. Helena Volk, Saša Piskernik, Marija Kurinčič, Anja Klančnik, Nataša Toplak, Barbara Jeršek. Journal of Food and nutrition Research (ISSN 1336-8672). Journal of Food & Nutrition Research, 2014, 53 (2): 97. Significant associations between increased skin thickness at the aPPD site and increased MAP ELISA S/P ratios were identified at each test time point. This was most notable on day 10 post-SICCT, where each 1mm increase in skin thickness was associated with an increase in S/P ratio of 4.4 ($p< 0.001$)
Conclusion : A significant association between aPPD hypersensitivity and MAP ELISA S/P readings was identified. Further research is required to identify whether utilizing aPPD measurements may identify animals that have previously been exposed to MAP, or alternatively, if MAP antibody ELISA is yielding non-specific responses to environmental M. avium species.

Keywords : DNA extraction, MAP, milk and colostrum, buffalo and bovine

P-03.4

Mycobacterium avium subsp. paratuberculosis isolation and PCR Real Time biomolecolar investigation on compost barn from a low prevalence dairy farm in the province of Rome

Gamberale F.[1], Pietrella G.[1], Maurelli M.[1], Scaramella P.[1], Maggiori F.[1], Di Domenico E.[1], Sagrafoli D.[1], Giacinti P.1, Cersini A.[1]

[1]*Istituto Zooprofilattico Sperimentale del Lazio e Toscana "M. Aleandri" Rome, Italy.*

In Italy, the number to biogas-fuelled dairy farms using solid separation bedding as compost barn, are constantly increasing. Such cattle hausing is widespread in Israel, North America and North Europe; it improves animal wellness, increasing both the milk production and the reproductive performace thus reducing the management costs. At present, further study is needed about the influence of the solid separation bedding, the compost barn, on the gastroenteritis propagation risk caused by Mycobacterium avium subsp. paratuberculosis (MAP) and on the udder health. However, several studies report the MAP resistance in the environment, manure and in row or pasteurised milk. The aim of the present study was to evaluate the role of this type of bedding as a possible risk factor for MAP persistence and spread, in a dairy farm with a low MAP prevalence (0,59% con IC 95% 0,48% - 0,77%) and equipped with a biogas system. Fourteen compost barn pools from as many boxes have been collected ; the material has been subjected to culture isolation and PCR Real Time IS900, specific for MAP. As a preliminary result, three Map strains have been isolated from three pools, and the presence of MAP DNA by PCR Real Time has been detected from seven pools. References1. Sukhbir K. Grewal, Sreekumari Rajeev, Srinand Sreevatsan and frederick C. Michel Jr."Persitence of Mycobacterium avium subsp. paratuberculosis and other zoonotic pathogens during simulated compostingm manure packing, and liquid storage of dairy manure" Appl. Environ. Microbiol. 2006, 72(1): 565, DOI: 10.1128/AEM.72.1.565-574.2006.2. Pier Giorgio Ventura Separato solido di liquami una lettiera " alternativa" supplemento informatore agrario pag. 31-33,29/2013.3. Regolamento (CE) n 1069/2009 del parlamento europeo e del consiglio del 21 ottobre 2009recante norme sanitarie relative ai sottoprodi di origine animale e ai prodotti derivati non destinati al consumo umano e che abroga il regolamento (CE) n. 1774/2002 .

Keywords : MAP, isolation,PCR real time, compost barn.

P-03.5

The diagnostic possibilities of bovine paratuberculosis using skin test with avian PPD

Alvarado Pinedo MF[1], Fernández M[1], Moyano RD[2]; Peralta LM[1], Di Paolo LA[1]; Romero MA[1], Romano MI[2], Travería GE[1]

[1]*Centro de Diagnóstico e Investigaciones Veterinarias (CEDIVE), Facultad de Ciencias Veterinarias. Universidad Nacional de La Plata, Buenos Aires, Argentina.* [2]*Instituto de Biotecnología, CICVyA INTA, Buenos Aires, Argentina*

Paratuberculosis an ileocolitis mostly present in ruminants is caused by Mycobacterium avium subsp paratuberculosis the clinical manifestations includes diarrhea, progressive weight loss, general wasting and decreased milk production. Cell-mediated immune response predominates earlier in the first stages of the disease, to achieve an early ante-mortem detection under field condition the skin test is one choice. The aim of this work was to set up some performance characteristics of skin test with avian tuberculin. No significant differences were found, with receiver operating characteristic curve (ROC) (P= 0,14) between cervical tuberculin skin test (CTST) and caudal fold tuberculin skin test (CFTST), although CTST detected more reactants. Bayesian approach found differences (P= 0,0031) between the means of two different concentrations of avian PPD (0,5 mg/mL vs. 2 mg/mL), showing higher skin fold differences with 2 mg/mL, however with ROC no differences were evident (P= 0,931) between both concentration AUC= 0,718 (0,5 mg/mL) and AUC= 0,719 (2 mg/mL). The ROC test estimate a sensitivity of 69% and a specificity of 86,2% and cut off > 0 mm. Bayesian inference calculate 42,5% for positive predictive value (PPV), 91,6% negative predictive value (NPV), 3,3 for positive likelihood ratio (LR+) and 0,4 negative likelihood ratio (LR-). Also, we apply the ROCRegression package to know the effect of covariates on the skin test, according to results, the performance of the test is improved if 3 to 5 rounds of serial skin tests is applied between an elapsed time not superior to 3 months in order to detect the higher number of subclinical diseased animals.

Keywords :
Paratuberculosis, diagnosis, avian tuberculin, bovine

P-03.6

Within-herd prevalence limits for the identification of paratuberculosis affected herds using antibody detection in pooled milk samples

Heike Köhler[1], Mario Ziller[2], Franziska Gierke[1], Friedrich-Loeffler[3], Karsten Donat[4]

[1]*Institute of Molecular Pathogenesis, Friedrich-Loeffler-Institut, Federal Research Institute for Animal Health, Jena, Germany* [2]*Biomathematics Group, Friedrich-Loeffler-Institut, Federal Research Institute for Animal Health, Insel Riems, Germany.* [3]*Institute Germany and Veterinary and Food Surveillance Office of the County Cloppenburg, Cloppenburg, Lower Saxony, Germany.* [1]*Institute of Molecular Pathogenesis, Friedrich-Loeffler-Institut, Federal Research Institute for Animal Health, Jena, Germany.* [4]*Animal Health Service, Thuringian Animal Diseases Fund (public-law institution), Jena, Thuringia, Germany*

Identification of Mycobacterium avium subsp. paratuberculosis (MAP) positive herds isthe first step in paratuberculosis control. Due to insufficient data about thediagnostic performance, pool milk sampling, being a widely used, cost-effective approach for the surveillance of infectious diseases in dairy cattle in general,has not been exploited for this purpose to a large extent. In regions with ahigh herd level prevalence, utilization of bulk or pool milk sampling may be consideredthe first step to identify the most affected herds within a voluntary controlprogram. Using a logistic regression model, in this study threshold levels ofthe apparent within-herd prevalence (WHPapp) were determined for theidentification of paratuberculosis positive herds by antibody detection in milkpools. To simulate bulk milks from different herds, a total of 74 milk poolswere prepared, 50 from two MAP-positive and 24 from one MAP-non-suspect herds,by random selection and pooling of up to 50 individual milk samples. Fourdifferent commercial ELISAs were used. For each pool considered equivalent toa herd, WHPapp was estimated independently for each test methodapplied (antibody detection in milk or serum samples or faecal culture). Forpool milk testing, cut-off values of the 4 ELISAs were revised to ensure 99%specificity and high sensitivity as well, resulting in lower values than forindividual milk samples. WHPapp estimates based on antibodydetection tended to under-estimate the true situation compared to individualfaecal culture (FWHPapp). The estimated WHPapp thresholdsof all 4 pool-milk ELISAs increased with ascending probability of detection. For 50% probability of detection, FWHPapp thresholds of 8.9-16.4%were determined, increasing to 20.0-37.8% for 95% probability of detection. Theresults underline that antibody detection in pool milk allows only theidentification of herds with a very high prevalence of MAP shedders. In controlprograms it can be the first step to identify the most affected herds. However,it is unsuitable for prevalence investigations and monitoring.

Keywords :
Mycobacterium avium subsp. paratuberculosis, herd level diagnosis, faecal culture, enzyme linked immunosorbent assay (ELISA), cut-off values

P-03.7

MULTIPLEX DETECTION OF BOVINE ANTI-MAP ANTIBODIES BY LUMINEX SUSPENSION ARRAY

Michele Gastaldelli[1], Elisabetta Stefani[1], Stefania Fontana[2], Beatrice Boniotti[2], Ad Koets[3], Heike Koehler[4], Maria Lodovica Pacciarini[2], **Nicola Pozzato**[1]

[1]*Istituto Zooprofilattico Sperimentale delle Venezie, Sezione di Verona, Verona, Italy.* [2]*Istituto Zooprofilattico Sperimentale della Lombardia ed Emilia Romagna, Reparto Genomica, Brescia, Italy.* [3]*Central Veterinary Institute (CVI) part of Wageningen UR, Lelystad, The Netherlands.* [4]*Friedrich-Loeffler-Institut, Institute of Molecular Pathogenesis, Jena, Germany*

ELISA serological tests represent an important tool in Johne's disease (JD) control programs. The Luminex suspension array (Luminex) is a technology platform that allows the simultaneous detection of antibodies against multiple antigen targets. In this study we tested several antigens (3 purified protein derivatives Johnei (PPD-J) preparations, 4 Escherichia coli recombinant Mycobacterium avium ssp. paratuberculosis (MAP) proteins and one MAP peptide) to be applied in this platform. We coupled each antigen with a bead set via a carbodiimide reaction and applied the protocol provided by Luminex with the following modifications and settings. Serum samples were pre-absorbed with Mycobacterium phlei to remove non-specific reactions, plate were read by "high" acquisition mode using the Bioplex TM200 (Bio-Rad) and results were expressed as net median fluorescence intensity of 2 replicates.Due to technical issues with some antigens (high background or ineffectiveness) we eventually optimized the JD-Luminex test with 3 recombinant proteins: Ag1 (MAP0210c), Ag2 (MAP2942), Ag7 (MAP2609), all produced at CVI.For test evaluation we analysed 737 sera (510 JD negative from uninfected herds and 227 MAP culture positive of which 165 sera tested positive in ELISA). Cut-off values and Se of each antigen were defined applying ROC analysis and fixing the Sp of each antigen >99%. The performance of the JD-Luminex test was calculated with the cumulative result of Ag1, Ag2 and Ag7, considering at least one positive antigen (Se=23.8% and Sp=99.4%). The parallel application of a commercial ELISA (IDEXX) demonstrated a good agreement with the JD-Luminex test with a comparable Se (23.8 vs 27.3%).We further examined 189 sera from different Tuberculosis (TB) herd status:ELISA positive/ MAP culture negative from TB free/JD infectedJD vaccinated (Silirum-CZV) from TB free/JD infectedTB culture positive cattle from TB infected/JD unknownOther cattle from TB infected/JD unknownThese results demonstrated no interference in sera from TB infected and JD vaccinated cattle.To resume, we have standardized a JD-Luminex serological assay for cattle. The performance of the test is comparable to commercial ELISAs and can be easily implemented with additional antigens, when available, to improve the performance of the test.

Keywords :
Suspension array, serology, multiplex

P-03.8

ELISA TEST FOR PARATUBERCULOSIS ANTIBODIES IN BOVINE BULK MILK, NAÏVE OR CONCENTRATED : A FIELD EVOLUTION

Laura Lucchese[1], Costanzi Carlo[2], Dorigoni Vittorio[2], Moresco Angela[3], Dellamaria Debora[1], Nardelli Stefano[1]

[1]*Istituto Zooprofilattico Sperimentale delle Venezie, Legnaro PD, Italy .*[2]*Servizio Politiche Sanitarie e per la Non Autosufficienza, Provincia Autonoma, Trento, Italy.* [3]*Azienda Provinciale per i Servizi Sanitari, Trento, Italy*

Italian guidelines for paratuberculosis control, issued in 2013, define different qualifications, basing on clinical and serological findings. Guidelines acknowledge the first accreditation level (PT1) to farms having a low (ffi5%) seroprevalence, measured in a random sample of cows aged >36 months. Antibodies are detected by means of ELISA carried out on individual (blood or milk) samples. Bulk milk testing, much more convenient, is not taken into account due to its lower sensitivity. The concentration of the immunoglobulins G (IgG) was evaluated as a pre-treatment for the bulk milk serology, aiming at identifying high-prevalence (>5%) farms.During the surveillance carried out in the province of Trento (Italy) in 2015, raw bulk milk samples (RBMS) were collected in 250 dairy farms together with blood samples from all cows aged >36 months: positive farms (i.e. having at least one positive cow) were 64, out of which 30 showed a high (>5%) seroprevalence.10 ml of RBMS were concentrated with Eradikit® Milk IgG Purification Kit (In3 Diagnostic): after casein precipitation, IgG were captured using an affinity matrix and resuspended to a 275 µl volume (concentrated bulk milk sample – CBMS). RBMS and CBMS were analyzed with ELISA IdScreen® Paratuberculosis Indirect (IDVet, confirmation format). For CBMS, a higher sample dilution (1:4) during initial absorption phase and a higher positive cut-off value (S/P>0,20) were chosen for ELISA test.CBMS ELISA provided a specificity of 91% and a sensitivity of 44% / 70% (in low / high-prevalence farms respectively), whereas RBMS ELISA gave a 99% specificity and 15% / 60% sensitivity. Concentration effect is more visible in low-prevalence farms. All 9 high-prevalence farms having a CBMS false-negative result were resampled: 5 of them turned positive in CBMS ELISA, thus increasing the overall sensitivity to 87%. The 17 CBMS false-negative farms were also retested, and 13 of them turned negative in CBMS ELISA.Repeated bulk milk sampling actions increases sensitivity of CBMS ELISA to values which could be considered adequate for achieving the PT1 accreditation level. However, the CBMS ELISA procedure suffers from the detection of some false-positives, which requires an improvement of its specificity.

Keywords :
Paratuberculosis, bulk milk, IgG purification, ELISA

P-03.9

Volatile organic compounds from Mycobacteriumavium ssp. paratuberculosis indicate bacterial growth - a new accelerated diagnostic approach ?

Anne Küntzel[1], Sina Fischer[1], Andreas Bergmann[2], Peter Oertel[2], Marcus Steffens[2], Phillip Trefz[2], Wolfram Miekisch[2], Jochen Schubert[2], Petra Reinhold[1], Heike Köhler[1]

[1]*Institute of Molecular Pathogenesis at the 'Friedrich-Loeffler-Institut', Jena, Germany.* [2]*Department of Anesthesia and Intensive Care, University of Rostock, Germany*

Background: Detection of volatile organic compounds (VOCs) released into gas phases above bacterial cultures is currently discussed as new and promising diagnostic approach. The aim of this study was to evaluate the VOC profile linked to growth of Mycobacterium avium subsp. paratuberculosis (MAP) with respect to different strains, different bacterial counts, and different periods of incubation. Animals and Methods: VOCs were measured above 3 different MAP strains (2 MAP Type-II, 1 MAP Type-III) cultivated on Herrold's Egg Yolk Medium (HEYM) and above medium slants not inoculated with bacteria as controls. The vials were inoculated with 4 different bacterial counts and cultivated for 2, 4, and 6 weeks. For pre-concentration of VOCs, needle-trap microextraction (NTME) was used. All VOC analyses were performed using gas chromatography / mass spectrometry. Results: Altogether 68 VOCs could be measured in the headspace above MAP-inoculated and control slants. More than 25 VOCs were identified as potential biomarkers of MAP growth. The latter differed significantly from pure media and were significantly higher concentrated above MAP cultures compared to the surrounding laboratory room air. The VOC profile was defined after 6 weeks of cultural incubation assuming that the bacteria were in exponential growth at that time. These substances belonged to the classes of alcohols, aldehydes, esters, ketones, aliphatic and aromatic carbohydrates. Already after 4 weeks of incubation, about 15 VOCs differed significantly from the control vials. These results were valid for all three MAP strains. Most VOC concentrations were directly related to increasing bacterial counts, either by initial inoculum or by bacterial growth over time. Others were either reaching a peak and decreasing then or VOC concentrations were decreasing while control vials showed higher concentrations. Remarkable time-dependent changes of the VOC concentrations in the headspace above control slants were also noted. Conclusions: Data support the hypotheses that (i) VOCs provide information about growth of MAP strains and (ii) cultural diagnosis could be accelerated by taking specific VOC profiles into account. Furthermore, the results stress the importance of a profound knowledge about influences on VOC composition before defining reliable and accurate marker sets for diagnostic purposes for any bacteria.

Keywords : VOC, kinetics, cultures, bacterial counts, GC-MS

P-03.10

Comparative Study on the PMS-Phage Assay and qPCR for Detection and Enumeration of Mycobacterium avium subsp. paratuberculosis in Milk in Two Laboratories

S. Butot[1], M. Peterz[1], L. Michot[1], V. Sonnard[1], B. Jagadeesan[1], V. Bastic Schmid[1], I.R. Grant[2], J. Donaghy[3]

[1]*Nestlé Research Center, Nestec Ltd, Vers-chez-les-Blanc, CH-1000 Lausanne, Switzerland.* [2]*Institute for Global Food Security, School of Biological Sciences, Queen's University Belfast, Medical Biology Centre, 97 Lisburn Road, Belfast BT9 7BL, Northern Ireland, United Kingdom.* [3]*Nestlé S.A., Avenue Nestlé 55, CH-1800 Vevey, Switzerland*

Introduction: Monitoring of milk supplies for the presence of Mycobacterium avium subsp. paratuberculosis (MAP) merits attention by the dairy industry. A peptide-mediated magnetic separation (PMS)-phage assay for the enumeration of viable MAP in milk has been recently developed which may have application in this context. So this study was carried out to determine whether the PMS-phage assay could be successfully transferred from a university laboratory to a dairy industry laboratory.Materials and Methods: Two comparative trials were undertaken between the two laboratories (A and B), during which milk or milk product samples spiked with varying levels of MAP were blind-tested using the PMS-phage assay. Samples were also tested in parallel by two qPCR methods, to assist with interpretation of results. Results: In the first trial, the PMS-phage assay performed as expected when carried out in Laboratory A; quantitative results obtained correctly reflected the number of MAP inoculated into raw milk. However, reduced detection sensitivity, inconsistent plaque counts, and a background flora problem were encountered when performed in Laboratory B. In contrast, a PMS-thermal step qPCR assay and a commercial Adiapure-qPCR assay, applied in Laboratory B, gave results consistent with the original MAP spiking levels in raw milk. In a second trial, results produced with the PMS-phage assay applied by an analyst from Laboratory A in Laboratory B were also inconsistent, indicating a limited detection of viable MAP from all the milk matrices (raw whole, raw skimmed, UHT whole, and powdered milk) and broth analysed. Both apparent false positive and false negative PMS-phage assay results were obtained. In contrast, during both trials the Adiapure-qPCR results obtained by Laboratory B were consistent with MAP spiking levels. Conclusions: Significant problems were encountered when the PMS-phage assay was transferred from Laboratory A to Laboratory B, the causes of which were not fully elucidated. They could potentially be analyst-related or due to differences in equipment employed. On the basis of the results and experiences described, it was concluded that in its present multi-step format, the PMS-phage assay does not represent a suitable test for implementation by dairy laboratories.

Keywords : Mycobacterium avium subsp. paratuberculosis, milk, PMS-phage assay, PCR, method transfer

P-03.11

Italian Johnin in the Interferon Gamma Release Assay for Ovine Paratuberculosis diagnosis: preliminary data

Di Paolo A.[1], **Mazzone P.**[1], Broccatelli S.[1], Cordovani E.[1], Costarelli S.[1], Caporali A.[1], Gobbi M[1], Corneli S.[1], Biagetti M.[1], Ciullo M.[1], Pezzotti G.[1], Papa P.[1]

[1]*Istituto Zooprofilattico Sperimentale dell'Umbria e delle Marche, Via G. Salvemini 1, 06100 Perugia, Italy*

The diagnosis of Paratuberculosis (PTB), due to Mycobacterium avium subsp. paratuberculosis (MAP), is usually based on serology, PCR and fecal culture. These traditional tests detect animals only in advanced stages of infection, when they are shedding MAP; therefore there is great interest in developing tests which act earlier, when cell-mediated immune (CMI) response predominates.Interferon Gamma (IFN-γ) Release Assay (IGRA) highlights CMI response, quantifying the cytokine produced by lymphocytes stimulated with proteins extracted from Mycobacteria. The assay has been studied extensively and optimized for PTB immune-diagnosisin cattle, while few studies have been performed with the same purpose for ovine.Aimed to assess the diagnostic ability of IFN-γ test in detecting MAP-infected sheep at the beginning of the infection, we performed a IFN-γ test with an Italian Johnin protein purified derivative (PPDJ), extracted from a field MAP strain, already used in bovine.Briefly, 3 flocks (one PTB-positive, one historically PTB-negative, one with a PTB control program ongoing) from Central Italy were included in the study (46 sheep born between 2010 and 2013), all animals were evaluated for PTB status by serum ELISA and PCR from faeces. Whole blood aliquots were stimulated with: PBS (Nihil), Bovine PPD, Avian PPD, PPDJ at two different dilutions (1:5; 1:10) and Pokeweed mitogen. IFN-γ production was assessed with Bovigam® kit.The positivity was assigned with 4 different interpretative criteria based on values recorded at 450nm. Among 22 ELISA positive sheep (47.82%), only 8 were identified as MAP infected in the IFN-γ test; this low number is probably due to the age of animals and the advanced stage of PTB. Among 24 ELISA negative (52.17%), 4 animals, from the flock with a PTB control plan, reacted to Johnin, in all criteria; they could be early infected sheep to be followed up. At the moment further investigations and appropriate interpretative criteria should be applied in ovine. Additional data may support the utility of IFN-γ test in the management of PTB infection in the flock, especially among young animals and in different epidemiological realities.Granted by Italian Ministry of Health RCIZSUM 04/2013.

Keywords :
Paratuberculosis, Johnin, Interferon-Gamma assay, sheep

P-03.12

Surveillance for clinical paratuberculosis in cattle in Finland

Seppänen Jaana[1], Skrzypczak Teresa[1], Pelkola Kirsti[1], Pelkonen Sinikka[1]

[1]*Finnish Food Safety Authority Evira, Finland*

In Finland, symptomatic paratuberculosis was recorded for the first time in 1992 since the early 1900s. During 1992 - 2000 paratuberculosis was detected in five beef herds. It has never been found in dairy cattle in Finland. The disease is not controlled by law. However, the Animal Health ETT advices to import only animals that have been tested and origin from paratuberculosis free herds. Since 2005, there has been a symptoms-based passive surveillance program for paratuberculosis in cattle. This includes free testing of samples from cattle presenting signs of paratuberculosis, typical lesions at slaughter or autopsy, or positive at antibody testing for various reasons. The program is run by the Finnish Food Safety Authority Evira. The veterinarians and farmers are encouraged to submit samples. Faecal and organ samples are Ziehl Neelsen stained and cultured for Mycobacterium avium subspecies paratuberculosis, and serum samples tested for specific antibodies with ELISA.The aim is to assess the presence of paratuberculosis and the need of a specific control program.Samples for bacteriological culture were received from 111 animals during 2005 – 2015. Eightysix animals from 42 herds had clinical symptoms or typical lesions and remaining 25 animals had been positive in previous antibody testing. Serum samples were received from 47 of those animals with symptoms or typical lesions. Two of these samples were positive with ELISA. No M. avium ssp. paratuberculosis bacteria were isolated from the faecal or organ samples.To conclude, there has not been any confirmed paratuberculosis in cattle in Finland since the year 2000.

Keywords :
Paratuberculosis, cattle, Finland

P-03.13

Identification of Immuno-reactive secretory proteins of Mycobacterium avium subspecies paratuberculosis vaccine strain 'S 5' with a diagnostic potential in early infection

Saurabh Gupta[1,2], Kundan Kumar Chaubey[1,2], A.K. Bhatia[1], **Shoor Vir Singh**[2], Naveen Kumar[3], Sahzad[2], Manju Singh[2], Bjorn John Stephen[2], Sujata Jayaraman[4], Mukta Jain[4], Jagdip Singh Sohal[4]

[1]*Department of Biotechnology, Microbiology and Immunology, GLA University, Mathura, Uttar Pradesh, India.* [2]*Animal Health Division, Central Institute for Research on Goats, Makhdoom, PO-Farah, Mathura, Uttar Pradesh, India.* [3]*Veterinary Type Culture Collection, NRC on Equines, Indian Council of Agricultural Research, Hisar, Haryana, India.* [4]*Amity Institute of Microbial Technology, Amity University Rajasthan, Kant Kalwar, Jaipur, Rajasthan, India.*

Johne's disease, chronic enteritis of ruminants caused by Mycobacterium avium subspecies paratuberculosis (MAP), is responsible for extensive economic losses to animal farmers and dairy industry worldwide. Current diagnostics for chronic disease like JD are hampered by the lack of specific immune-reactive antigens present in local strains. The capability of diagnosing the disease rapidly and identifying its causative agent quickly and correctly is critical to combat this disease and halt the progression into the epidemics like situation. Cross-reactivity of the antigens and poor sensitivity and specificity of the current diagnostics is a problem due to the extensive sharing of antigenic epitopes by MAP and other mycobacterial species. Present study is focused on the search of new candidate antigenic epitopes to be evaluated as novel biomarkers for early ParaTB diagnoses. Culture filtrate (CF) proteins profile of vaccine novel strain 'S 5' of MAP (Indian Bison Type) and their immuno-reactivity was studied in the early growth period (4 and 6 weeks) in Middlebrook 7H9 medium supplemented with ADC, PANTA antibiotics and mycobactin J. Analysis of harvested CF proteins was done by sodium dodecyl sulfate polyacrylamide gel electrophoresis (SDS-PAGE). Immunoblotting showed reactivity of four CF proteins commonly recognized (19, 36-38 and 65 kDa) and an additional 48 kDa protein was recognized at 6 week by MAP infected caprine sera. Fractionation of CF proteins (4 and 6 weeks) was done by BioLogic LP chromatography system (Biorad) using Bio-ScaleTM Macroprep(R) High Q column. Combined application of these protein fractions were evaluated by Indirect ELISA in search of potential antigenic epitopes. Results showed the improved sensitivity and high specificity as judged by Receiver-operating curve (ROC) analysis as compared with whole cell sonicated semi-purified protoplasmic antigen (sPPA). We anticipated that the indigenously developed assays using immuno-reactive protein(s) of native vaccine strain will serve as backbone of future control programs in the country. Our studies have already shown that there cannot be universally efficient diagnostic kits. One should try to use antigens from local strains. By using universal commercial kits we are unknowingly under reporting the disease incidence.

Keywords :
Mycobacterium avium subspecies paratuberculosis, CF proteins, SDS-PAGE, Immunoblotting, Indirect ELISA

P-03.14

Comparison of 'Indigenous ELISA kit' with 'Ethanol Vortex ELISA kit (USA)' for detection of Mycobacterium avium paratuberculosis infection in cattle population in search of 'globally relevant kit'

Kundan Kumar Chaubey[1], Saurabh Gupta[1], Jagdip Singh Sohal[2], Sujata Jayaraman[2], Bjorn John Stephan[1], Manju Singh[1], Mukta Jain[2], Naveen Kumar[3], **Shoor Vir Singh**[1]

[1]Microbiology laboratory, Animal Health Division, Central Institute for Research on Goats, Makhdoom, PO-Farah, Mathura, 281122, India. [2]Amity Institute of Microbial Technology, Amity University Rajasthan, Kant Kalwar, NH 11C Delhi-Jaipur Highway, Jaipur, 303 002, Rajasthan. [3]Veterinary Type Culture Collection, National Research Centre on Equines, Hisar, Haryana-125001, India

Indigenous ELISA kit (India) was evaluated with Ethanol Vortex ELISA kit (USA) for diagnosis of MAP infection in cattle population. Of 160 cattle (118 Vaccinated and 42 non-vaccinated) were screened where, 129 (80.6%) and 35 (21.8%) cattle were positive by Indigenous ELISA and EV ELISA, respectively. Semi purified antigen used in Indigenous ELISA was harvested from native widely prevalent biotype of MAP (Indian Bison Type) in India. Indigenous ELISA exhibited higher sensitivity (91.4%) as compared to EV_ELISA. Use of semi-purified whole cell antigen mix obtained from strain 'S 5' of 'Indian Bison Type' biotype of goat origin was superior than using Ethanol Vortexed sub-species specific surface antigen from (Cattle Type) bio-type of 'Linda' strain of US of cattle origin in screening of domestic cattle population in India. This may be due to use of two strains from geographically juxtaposed countries like US and India. The study underlined that for the diagnosis of MAP infection, it will be prudent to use polyclonal antigens from locally available strains and it will not be worthwhile to invest time and energies in search of 'globally competitive kits' for the diagnosis of Johne's disease in local population.

Keywords :
Mycobacterium avium subspecies paratuberculosis, Indigenous ELISA, EV ELISA, Indian Bison type, Johne's disease

P-03.15

Potential and limitations of use of the Mycobacterium avium subsp. paratuberculosis specific antigen L5P and its variants produced by chemical synthesis

Sylvie Bay[1], Douglas Begg[2], Christelle Ganneau[1], Sébastien Holbert[3], Maxime Branger[3], Thierry Cochard[3], Marc Govaerts[4], Heike Koehler[5], Richard J. Whittington[4], **Franck Biet**[3]

[1]Institut Pasteur, Unité de Chimie des Biomolécules, Département de Biologie Structurale et Chimie, Paris, France, CNRS UMR 3523, Paris, France. [2]Faculty of Veterinary Science, University of Sydney, Camden, NSW, Australia. [3]INRA, UMR1282, Infectiologie et Santé Publique, INRA centre Val de Loire, F-37380 Nouzilly, France. [4]Operational Directorate Bacterial Diseases, Veterinary and Agrochemical Research Centre (CODA-CERVA), Groeselenberg 99, B-1180 Brussels, Belgium. [5]Friedrich-Loeffler-Institut, Federal Research Institute for Animal Health, Jena, Germany

Background: Unlike other MAC members, Mycobacterium avium subsp. paratuberculosis (MAP) does not produce GPL on the surface of the cell wall but a lipopentapeptide called L5P or ParaLP01. The molecular and genetic characterization of this antigen demonstrated that L5P is specific to MAP. Currently available diagnostic tests are based on the use of whole cell antigens that are mainly not MAP specific and that require a pre-absorption step with antigens of M. phlei. This diagnosis detect animals with a late stage infection but are not sensitive enough to detect early stage infection. We hypothesize that the pre-absorption step prevents detection of informative populations of immunoglobulins, especially in animals with subclinical faecal shedding of MAP that are not detected by the current serology-based diagnostics. The L5P antigen has now been used in various studies to evaluate a MAP specific sero-diagnostic. This synthetic molecule was used as a pure product, not requiring pre-absorption step. However a number of parameters must be assessed including the formulation of L5P and with reference to a panel of well defined serum samples.Objective: To assess the potential of L5P and its hydrosoluble variants for the serological diagnosis of MAP infection using collections of sera from different contexts.Method: In order to find the best compound for use in serology we chemically synthesized L5P and derivatives of L5P including water-soluble variants. These pure compounds were evaluated on collections of extensively characterized sera from infected and non-infected cattle, goats and sheep. In addition, the panels of sera included animals infected with M. bovis and M. avium subsp. hominissuis.Results: ROC analysis showed that L5P and also its water-soluble derivatives are suitable for the development of serological diagnosis. Advantageously, these pure synthetic MAP specific antigens can be produced at low cost. The use of L5P has not been validated in the contexts of ovine paratuberculosis. In the context of infections due to other mycobacteria such as M. bovis or the more closely related species M. avium subsp. hominissuis, the L5P did not cross react and therefore may be a valuable antigen to solve ambiguous results in other tests.

Keywords :
Lipopentapeptide, sero-diagnostic, antigen

P-03.16

Evaluation of the infection status of bovine animals positive on a MAP ELISA test following herd screening for Johne's disease in Ireland

Ramovic Elvira[1]

[1]*Central Veterinary Research Laboratory, Department of Agriculture Food and Marine, Backweston, Celbridge, Co. Kildare, Ireland*

The performance and application of diagnostic tests for Paratuberculosis (Johne's disease) is critical to both screening herds for infection and managing the infected herds. In recent years dairy herds in Ireland are participating in JD control primarily based on screening animals for MAP antibodies in milk or serum using an enzyme linked immunosorbent assay (ELISA).To assess current screening strategy, animals disclosed as positive on foot of an ELISA test were followed up with a faecal sample and where possible a repeat serum sample. Each herd owner was also requested to submit details of the previous ELISA test. Faeces were tested for the presence of MAP using TREK ESP II liquid culture system and Cornell double incubation decontamination method while the serum samples were tested using MAP ELISA (Idexx). The results of this study reflect a defined time period of one year.Samples were submitted from a total of 228 herds flagged as suspect for JD, and 11% were classed as infected. Of 828 individual faecal culture samples MAP was isolated from 65 animals (7.9%). Thirty three infected animals belonged to three herds. Of the animals tested in repeat ELISA 52.1% were negative, suggesting transient antibody responses may be occurring. Survival analysis of ELISA test positive animals continues.Annual tuberculin tests, climatic conditions and outdoor grazing for part of the year may impact on ELISA specificity in Ireland. If ELISA test is employed as a screening tool for MAP infection in herds of unknown status, faecal culture and repeat serum ELISA testing of seropositive animals is recommended.

Keywords :
Screening, ELISA, confirmation, culture, specificity

P-03.17

Potential of Mass Spectrometry for routine identification of Mycobacterium avium subsp. Paratuberculosis

Malia Yaici[1], Maria-Laura Boscirolі[2], Gilles Etienne[3], Franck Biet[4], **Jean-Louis Moyen**[1]

[1]*Laboratoire Départemental d'Analyse et de Recherche de la Dordogne (LDAR24), 24660 Coulounieix-Chamiers, France.* [2]*ANSES, Unité Zoonoses Bactériennes (UZB), Maisons-Alfort, France.* [3]*CNRS, Institut de Pharmacologie et de Biologie Structurale (IPBS), BP64182, F-31077 Toulouse, France.* [4]*INRA, UMR1282, Infectiologie et Santé Publique (ISP-311), F-37380 Nouzilly, France*

Mass spectrometry, increasingly used in laboratories for bacterial identification remains inadequate for identification of Mycobacterium avium subsp. paratuberculosis (MAP) and mycobacteria in general. Mass spectrometry may allow rapid differential diagnosis of mycobacterial distinguishing members of Mycobacterium tuberculosis complex, Mycobacterium avium Complex (MAC) and environmental mycobacteria.However, no routine clinical identification methods have been adapted to MAP that required both time consuming to obtain enough biomass for analysis. In addition the protein extraction protocols commonly used are not satisfactory and are labor intensive. Finally, the database containing spectra does not allow identification of mycobacteria to the subspecies level.The aim of this work is to develop methods of identification at the species and subspecies level for stains of MAC (M. avium sp avium, M. intracellulare, M. silvaticum and M. homnissuis) by Matrix-Assisted Laser Desorption-Ionization Time of Flight Mass Spectrometry (Maldi-tof MS) based on Vitek MS (Biomérieux) system, compare the sensitivity and the response time of the culture in liquid medium (7H9 and VersaTREK para-JEM) with solid medium method (Herrold agar), evaluate the performance of the identification of MAC mycobacteria in feces samples cattle and extend the spectra obtained for each strain studied so as to enrich the RUO (Research Use Only) (version 4.12) and IVD (In Vitro Diagnostic) (version 2.1.1) databases.Analysis of pure strains has led to identification at the complex, Mycobacterium genus and species level. Tests on sample cattle feces were also made.

Keywords :
Mycobacteria, Mycobacterium avium subsp. paratuberculosis, mass spectrometry, diagnosis

P-03.18

COMPARISON OF COMMERCIAL DNA EXTRACTION AND QPCR SYSTEMS FOR A BETTER SENSITIVITY IN DETECTING THE CAUSATIVE PARATUBERCULOSIS PATHOGEN IN DAIRY COW FECAL SAMPLES

D. Fock-Chow-Tho[1,2], E. Topp[3], and **N. Bissonnette**[1]

[1]Agriculture and Agri-Food Canada, Dairy & Swine Research and Development Centre2000 College Street, Sherbrooke, Quebec, J1M 0C8 Canada. [2]Department of Biology, Université de Sherbrooke, Sherbrooke, Quebec, J1K 2R1 Canada. [3]Agriculture and Agri-Food Canada, London, Ontario, N5V 4T3 Canada

Mycobacterium avium ssp. paratuberculosis (MAP) is the pathogen inducing ruminant paratuberculosis (Johne's disease) worldwide. While formerly of sporadic incidence, the prevalence of infected animals resulting from modern farming has raised paratuberculosis bovine disease to a global concern. Oral-fecal contamination is the most important mode of transmission of paratuberculosis. Hence, eradicating shedders could prevent MAP propagation. Whereas considered the standard method for MAP diagnosis, fecal culture requires specialised costly media and a long incubation time which sometimes resolve into disappointing bacterial contamination. To facilitate the efforts of control programs we evaluated the performance of direct fecal QPCR assays in terms of sensitivity. Several commercial kits use different strategies for DNA extraction and QPCR systems for capturing the presence of MAP in fecal samples. In this study, our aim is to compare the sensibility of detection of three commercially available DNA extraction kits broadly used in Canada, named A, B, and C, combined with two methods of QPCR detection (T and V). Forty-nine dairy cows from five different herds were sampled and diagnosed by fecal culture and ELISA assays to 6 months interval during two years. Their fecal samples were then tested eight times (8 replicates) with the respective DNA extraction method. While all of the three commercial DNA extraction kits were described as very efficient for the paratuberculosis diagnosis, method B allows a more sensitive detection than the two others. Indeed, 100% of cows declared positive for paratuberculosis by both fecal culture and ELISA assays were identified with method B while only 23% and 43% cows were confirmed with methods A and C, respectively. Interestingly, by using method B, low MAP shedders were detected. Moreover the QPCR system plays a critical role, with detection system T yielding QPCR reactions with the highest sensitivity. The results presented herein suggest that DNA extraction kit C in combination with QPCR system T allow successful amplification of MAP DNA from fecal samples with the highest sensitivity and specificity. The current study demonstrates the importance of testing different kits for DNA extraction from fecal samples and the impact of a QPCR system to identify MAP shedding animals.

Keywords :
Diagnosis, commercial DNA extraction system, fecal sample, quantitative polymerase chain reaction, sensitivity

P-03.19

PERIPHERAL PRODUCTION OF IFN-γ IN GOATS VACCINATED AGAINST PARATUBERCULOSIS IN RESPONSE TO DIFFERENT MYCOBACTERIUM AVIUM SUBSP PARATUBERCULOSIS (MAP) AND M. BOVIS ANTIGENS.

Royo M[1], Fuertes M[1], Fernández M[1], Castaño P[1], Ferreras MC[1], Benavides J[1], Pérez V[1]

[1]Dpt Sanidad Animal, Instituto de Ganadería de Montaña (CSIC-ULE), Facultad de Veterinaria, Universidad de León, León, Spain

Animals vaccinated against paratuberculosis could cross react in immune-based diagnostic methods for other mycobacterial diseases, mainly Mycobacterium bovis infections. The interference with the diagnostic tests used in tuberculosis eradication programs is the main reason barring the use of paratuberculosis vaccines in cattle in most countries. In this study, different antigens have been evaluated for immune-based diagnostic tests applied to goats vaccinated against paratuberculosis in order to evaluate whether they could be used to avoid cross reactions. Goats vaccinated at 1.5, 5 months old or when adults, were sampled at 4, 9 and 15 months post-vaccination (mpv) and whole blood was used for IFN-γ release assay. Paired unvaccinated controls were used in each group. For the assays, the following antigens were employed: avian and bovine PPD (CZ Veterinaria), johnin (Neiker), EC and HP M. bovis protein cocktails (Prionics,) and VK055 and VK067 Map proteins (Vacunek). No significant differences in the IFN-γ release was observed between vaccinated and unvaccinated animals in samples incubated with EC and HP M. bovis proteins or with VK067 Map antigen. The highest responses were obtained when johnin and avian PPD were employed, and also with bovine PPD, although always lower than the previous two antigens. Only samples from vaccinated animals incubated with VK055 Map protein gave positive reactions, lower that those observed with johnin or avian PPD. These results suggest that EC or HP M. bovis protein cocktails, already used for tuberculosis diagnosis, would be useful for discriminating paratuberculosis vaccinated from M. bovis infected animals.

Keywords :
IFN-gamma, Mbovis, vaccination, cross-reaction

P-03.20

Evaluation of different immunological methods for the diagnosis of paratuberculosis in dairy cows and their relationship with the fecal shedding determined by PCR

Fernández M[1], Mohamed Y[1], Fuertes M[1], Castaño P[1], Royo M[1], Ferreras MC[1], Benavides J[1], Pérez V.[1]

[1]Dpt Sanidad Animal, Instituto de Ganadería de Montaña (CSIC-ULE), Facultad de Veterinaria, Universidad de León, León, Spain.

Intra-herd prevalence of Mycobacterium avium subsp. paratuberculosis (Map) infection has been evaluated in three different cattle dairy herds, of 107, 64 and 34 adult animals respectively, using immunological diagnostic methods. Their efficiency in the diagnosis of the infection and the relationship of their results with the age of the animals were also assessed. Animals from the herds were examined at a single time point by serum indirect ELISA, using Map PPA-3 protoplasmic as coating antigen of the plates, and by of IFN-γ release assay in peripheral blood after incubation with avian PPD. Besides, the presence of specific antibodies in individual samples of milk and Map fecal excretion, evaluated by a real-time PCR (Paratb-kuanti VK), was assessed in all 34 animals from herd C. This study reflects a seroprevalence of 33.6%, 34.37% and 14.5% obtained in the three herds respectively, while 59%, 63.16% and 51.4% animals were positive to the IFN-γ release test. It was clear the significant relationship existing between the age of the animals and the response to each diagnostic test employed. In this sense, the percentage of ELISA positive animals was significantly higher in adult cattle than in younger animals. In contrast, the number of cattle positive to IFN-γ test was lower among adult cattle. In herd C, only 3 out of 34 cows were positive to ELISA in both milk and serum samples. Map fecal excretion was only found in the sole animal showing positive results to ELISA and IFN-γ test plus clinical signs of paratuberculosis. According to these results, the number of Map infected animals determined by immunological methods in a herd with paratuberculosis can reach high values. Response to immune-based diagnostic tests is related to the age of the animals and would precede fecal Map excretion.

Keywords :
Cattle, diagnosis, immune-based test, fecal excretion

P-03.021

Cell mediated immune response to L5P in longitudinal study of heifers from naturally Mycobacterium avium subsp paratuberculosis infected herd

S. Holbert[1], M. Branger[1], B. Lamoureux[2], L. Malherbe[2], C. Ganneau[3], T. Cochard[1], L. Olagnon[4], L. Comtet[4], C. Tholoniat[2], S. Bay[3], **F. Biet**[1]

[1]UMR1282, Infectiologie et Santé Publique, INRA Centre Val de Loire, University François Rabelais of Tours, F-37380 Nouzilly, France. [2]Groupement de Défense Sanitaire Région Centre-Val de Loire, F-36018 Châteauroux, France. . [3]Unité de Chimie des Biomolécules CNRS UMR 3523, Institut Pasteur, F-75015 Paris, France. [4]IDvet Laboratoire, 310 rue Louis Pasteur, 34790 Grabels

Peptidyl moiety of cell wall lipopentapeptide (L5P) specific of Mycobacterium avium subsp paratuberculosis (Map) is immunogenic and a target for specific humoral response in Map infected animals. A chemically synthesized L5P is able to induce specific cell mediated immune response (CMIR) in IFN-γ release assay (IGRA) in selected cows from Map infected herds comparatively to non-infected or M. bovis infected. Following these observations, the aim of this study was to evaluate if L5P was an antigen of early specific immune response and potentially a predictive tool of Map infection.
113 heifers of 6 herds were included in a two years' longitudinal study: 71 animals from three Map culture-confirmed herds, 11 animals from a Map infected herd Silirum® vaccinated during the study and 31 animals from two certified Map free herds. The analysis of the CMIR was investigated by IGRA following whole blood stimulation with synthetic L5P or mycobacterium purified protein derivative (PPD) from M. avium (PPDa), M. bovis (PPDb), Map (PPDj) and M. phlei (PPDp). Humoral immune response was quantified by L5P-based ELISA using an internal procedure and two commercial Map kits. Moreover, bacilli excretion was estimated by isolation and culture from faecal sample. PPDs' CMIR was more or less high depending of infected herds context, became high over 2 S/P ratio for 10/11 animals just after Silirum® vaccination and was low, less than 0.1 S/P ratio, in certified Map free herds. L5P CMIR was observed in 9 of 71 animals from Map culture-confirmed herds. These 9 animals with a L5P CMIR positive between 0.05 and 0.6 S/P ratio were from the same herd, knowing that L5P CMIR was previously detected in all included Map culture-confirmed herds. L5P CMIR was fluctuant as already described for PPD but was significantly correlated with PPDj CMIR. For 2 of the 9 animals, the L5P CMIR was predictive of the Map positive serology, whereas it was concomitant with seropositivity for 2 others and that for 5 animals was several times observed without seroconversion. And no seroconversion was observed in other herds. The continuation of this study would assess the predictive potential of L5P CMIR for paratuberculosis diagnosis.

Keywords :
Lipopentapeptide (L5P), Antigen specific, IGRA, CMI, Predictive diagnosis.

P-03.22

Molecular diagnosis of paratuberculosis (Mycobacterium avium subspecies paratuberculosis) and bovine tuberculosis (Mycobacterium tuberculosis Complex) by PCR using bovine blood samples

Park-de la Torriente, Alba[1*]; Maldonado-Castro, Edith[1]; Castellón-Ahumada, Victoria Elizabeth[1], González-Rodríguez, Diana[1]; **Chávez-Gris, Gilberto**[1*]

[1]*UNAM, FMVZ, CEIEPAA, Tequisquiapan, Queretaro, Mexico*

This project was supported by UNAM-DGAPA, PAPIIT IT202914 Paratuberculosis and bovine tuberculosis are diseases that cause an economic impact in the agricultural industry and potentially zoonotic. Paratuberculosis has no official eradication program even though it has a great distribution in Mexico. Diagnosis is important in order to establish a control of both diseases, is recommended that two or more diagnostic tests are used simultaneously. This work aims to develop a protocol for the use of PCR as a simultaneous diagnostic test to paratuberculosis and bovine tuberculosis from blood samples.
120 blood samples were taken from bovines from three different herds. The samples were used in three different trials: the first one was based on the amplification of a fragment from the IS900, the second was based on the amplification of the IS6110, the third one was a trial based on a Multiplex PCR for the simultaneous diagnosis of both mycobacteria. The results indicate that 0.83% (1/120) of the tested animals were positive to paratuberculosis using IS900 PCR, while 4.16% (5/120) of the animals were positive to bovine tuberculosis when tested with IS6110 PCR. Results from the Multiplex PCR showed that 0.83% (1/120) of the animals were positive to paratuberculosis. The difference in the results from the multiplex PCR and the IS PCRs could be due to the high specificity detected using multiplex PCR, that gave a lower sensibility; while the IS900 and IS6110 are highly conserved sequences which can also be found several times within the bacteria's genome. This could improve the sensitivity of the essay. This work shows that the simultaneous diagnosis of paratuberculosis and tuberculosis from blood samples is possible by PCR, however, it also shows the need of improving the sensitivity of the test without compromising its specificity. ReferencesSchiller I, Oesch B, Vordermeier HM, Palmer MV, Harris BN, Orloski KA, Buddle BM, Thacker TC, Lyashchenko KP, Waters WR, 2010. Bovine tuberculosis: A review of current and emerging diagnostic techniques in view of their relevance for disease control and eradication. Transboundary and Emerging Diseases. 57, 205–220. Ollar RA, Connell ND. 1999. Molecular Mycobacteriology: Techniques and Clinical Applications. New York, U.S.A.: Marcel Dekker.

Keywords :
PCR in blood, IS900, Paratuberculosis, tuberculosis

P-03.23

P35 recombinant protein, production and use in ELISA test for ovine paratuberculosis in Mexico

Castellón-Ahumada Victoria[1], González, Diana[1]; Park Alba[1]; Maldonado-Castro, Edith[1]; Chávez-Gris, Gilberto[1*]

[1]*UNAM, FMVZ, CEIEPAA, Tequisquiapan, Queretaro, Mexico*

The objective in this study was to produce a recombinant protein (P35) of Mycobacterium avium subsp. paratuberculosis (Map) by M13 phage display that could be use as antigen in ELISA. The P35 was identify by immundetection like Western Blot through E tag, in addition to monoclonal and polyclonal antibodies, the molecular weight of protein was always 40 kDa, because P35 weight 35 kDa and is attached to VII protein from phage which has a molecular weight of 5 kDa. The P35 was tested by enzyme linked immune sorbent assay (ELISA) with 48 sheep´s serological samples, organized in two groups, a positive with 36 and a negative with 12. The results showed that the ELISA using the P35 has a sensitivity of 100% and specificity of 92%.This project was supported by UNAM-DGAPA, PAPIIT IT202914 ReferencesMikkelsen H, Aagaard C, Nielsen SS, Jungersen G. Review of Mycobacterium avium subsp. paratuberculosis antigen candidates with diagnostic potential. Vet Microbiol. 2011;152(1-2):1-20. Galfrè G, Monaci P, Nicosia A, Luzzago A, Felici F, et al. Immunization with phage-displayed mimotopes. Methods Enzymol. 1996;267:109-15.

Keywords :
Map, P35, phage display, ELISA

P-03.24
Detection of Mycobacterium avium subspecies paratuberculosis in environmental samples

Orachun Hayakijkosol[1], Robert Hedlefs[1], Elizabeth Kulpa[1], Wayne Hein[2], Jacqueline Picard[1]

[1]*College of Public Health, Medical and Veterinary sciences, James Cook University, Townsville, Queensland, Australia.* [2]*School of Animal and Veterinary Sciences, The University of Adelaide, Adelaide, South Australia, Australia*

Mycobacterium avium subspecies paratuberculosis (MAP or Mptb) is the cause of Johne's disease in wild and domestic ruminants. Animals clinically affected by this disease become severely emaciated and develop projectile diarrhoea prior to death. Infection of ruminants is usually by the ingestion of contaminated material. The current global diagnostic standard is to culture faeces, but it can take up to three months to detect Mptb by the culture method. Recently Australia has included the rapid High Throughput Johne's test polymerase chain reaction (HT-J PCR) as the diagnostic standard for presence of the organism in faeces. However, the HT-J PCR requires expensive equipment and is best suited to large numbers of samples. This study aimed to develop a manual equivalent protocol to the HT-J PCR for DNA detection that is cheaper and suitable to process small number of samples. RNA and propridium monoazide (PMA) methods were used to confirm survivability of the bacteria in environmental samples. Soil samples from 4 different locations (Dingo, Malanda, Townsville and Charters Towers) of north Queensland, Australia and sterilised faeces mixed with 2 strains of Mptb (Bison and Bovine) were used in this study. The samples were placed in wet and dry conditions and left in their locations for 16 months. Each month, eight samples (2 samples each Mptb strain and 2 samples from each condition) were collected for DNA, RNA and PMA tests. DNA, RNA and PMA results showed that Mptb in wet condition started to decrease after 3 months of environmental exposure while some Mptb in dry condition are still able to be detected after 16 months of this study. In the wet condition, RNA and PMA detections determined poor survivability of Mptb compared to dry condition. PCR results showed that Bison and Bovine strains of Mptb samples had similar results in dry and wet conditions. PMA method determined that less than half of DNA positive results came from live bacteria. This study showed manual DNA technique can be used to detect Mptb in environmental samples and PMA can be used to determine survivability of Mptb. Also, the bacteria maybe live longer in dry environment.

Keywords :
Detection, Survivability, Mycobacterium avium subspecies paratuberculosis, Environmental samples

P-03.25
Optimization of marker assay for diagnosing the immune response due to vaccination and natural infection

Jagdip Singh Sohal[1], Sujata Jayaraman[1], Kundan Kumar Chauabey[2], Mukta Jain[1], Rinkoo Devi Gupta[3], Gajendra Kumar Aseri[1], Shoor Vir Singh[2], Saurabh Gupta[2], Manali Datta[4], Neelam Jain[4], Parul Yadav[1], Neeraj Khare[1]

[1&4]*AIMT & AIB, Amity University Rajasthan, Jaipur, India.* [2]*Animal Health Division, Central Institute for Research on Goats, Mathura, Uttar Pradesh, India.* [3]*South Asian University, New Delhi, India*

Paratuberculosis or Johne's disease (JD) caused by Mycobacterium avium subspecies paratuberculosis (MAP) results in huge economic losses to ruminant industry globally. It is chronic granulomatous enteritis leading to protein loosing enteropathy and only limited chemotherapeutic preparations have anti-MAP activity. MAP is not killed during pasteurization, therefore milk acts as an important vehicle for the distribution of bacilli to human beings and is now in all certainty the cause of Crohn's disease in humans. Therefore in recent past control of paratuberculosis in animals has become a priority due to farm economics and public health concerns. Unless infection is controlled in livestock population, bug will continue to be distributed through milk chains. Recent researches have proved that vaccination can be an effective and practical way to control this disease. Our group has developed highly efficacious indigenous therapeutic vaccine for treatment and prevention of paratuberculosis. However for field use of this effective vaccine, we need to develop assay that can distinguish between vaccinated and infected animals. We selected four antigens that are not part of the vaccine and developed an ELISA based assay that can easily discriminate the immune response due to vaccine or infection by simply converting the OD values to S/P ratio. The S/P ratio range observed for healthy, vaccinated and infected animals was; 0.00–0.09, 0.10–0.40 and >0.4–10.0, respectively. Therefore using 'single ELISA test' we can easily know the status of animal. Optimized test is under field testing and validation, and if properly used can serve as important tool in the control of this incurable disease of domestic livestock in the country and also boost per animal productivity.

Keywords :
Paratuberculosis, vaccination, natural infection, differentiation

P-03.26

THE CHALLENGE OF CONFIRMING JOHNE'S DISEASE: AN EVALUATION OF 14 CATTLE WITH SUSPECTED MYCOBACTERIUM AVIUM SUBSP. PARATUBERCULOSIS (MAP) INFECTION BASED ON HERD STATUS AND/OR SUGGESTIVE CLINICAL SIGNS

Louise Britton[1], Conor McAloon[1], Aideen Kennedy[2], Riona Sayers[2], Joseph Cassidy[1], Jim O'Donovan[3], Nola Leonard[1], Stephen V. Gordon[1], Bryan Markey[1]

[1]School of Veterinary Medicine, University College Dublin (UCD), Belfield, Dublin, Ireland. [2]Animal and Grassland Research and Innovation Centre, Teagasc, Moorepark, Fermoy, Cork, Ireland. [3]Department of Agriculture, Food and the Marine (DAFM), Regional Veterinary Laboratory, Model Farm Road, Cork, Ireland

In India, limited information is available on the presence of Mycobacterium avium subspecies paratuberculosis (MAP) infection (paratuberculosis) in farmer's herds of domestic livestock species. In present study, a total 420 fecal samples were collected from domestic animals (cattle-176, buffaloes-147, goat-79 and sheep-18) reared under semi- intensive farming conditions in farmer's herds of Uttar Pradesh and Madhya Pradesh states of North India. All the fecal samples were belonged to female animals. Of the 424 fecal samples, 345 and 75 samples were collected from different geographical regions of Uttar Pradesh and Madhya Pradesh states of North India, respectively. All the fecal samples were screened for the presence of MAP using microscopic examination and direct IS900 PCR. Of the 420 fecal samples, 90 (21.4%) and 47 (11.1%) fecal samples were positive for the presence of MAP using microscopic examination and direct IS900 PCR, respectively. Species wise, 37 (21.0%), 29 (19.7%), 23 (29.1%) 1 (5.5%) and 26 (14.7%), 11 (7.4%), 10 (12.6%), 0 (0.0%) samples were positive for the presence of MAP from cattle, buffaloes, goats, and sheep using microscopic examination and IS900 PCR, respectively. Presence of acid fast bacilli was higher in adult animals (21.8%) as compare to young animals (14.2%) using microscopic examination. The prevalence of MAP was higher in animals of Uttar Pradesh (microscopic examination-23.1%, direct IS900 PCR-11.5%) as compare to Madhya Pradesh (microscopic examination-13.3%, direct IS900 PCR-9.3%) of North India. Present study reported moderate prevalence MAP infection in farmer's herds of domestic livestock species and indicated the presence of other acid fast organisms in animals of Uttar Pradesh and Madhya Pradesh states of North India. Study highlights the need of surveillance and control programs for mycobacterial pathogens in order to reduce economic losses and protection of human heath in the country.

Acknowledgement: Authors are thankful to Indian Council of Agriculture Research, New Delhi for providing the financial support for this work.

Keywords : Case studies, post-mortem, MAP, differential diagnoses, inter-current disease

P-03.27

THE USE OF MALDI-TOF APPROACH FOR A RAPID IDENTIFICATION OF MAP: PRELIMINARY RESULTS

Ricchi Matteo[1], Mazzarelli Antonio[2], Di Caro Antonino[2], Savi Roberto[1], Leo Simone[1], Cammi Giuliana[1], Garbarino Chiara[1], Arrigoni Norma[1]

[1]Istituto Zooprofilattico Sperimentale della Lombardia e dell'Emilia Romagna, Centro di referenza Nazionale per la Paratubercolosi, Strada Faggiola 1, 29027, Gariga di Podenzano (PC), Italy. [2] Istituto Nazionale per le Malattie Infettive "L.Spallanzani", IRCCS, Laboratorio di Microbiologia e Banca Biologica, Via Portuense 292, 00149 Roma, Italy

Epidemiological tools applied in the epidemiology of paratuberculosis are based on DNA fingerprinting or whole genome sequencing of Mycobacterium avium subsp. paratuberculosis (MAP)[1, 2]. However, no method has so far investigated protein profiles useful for the identification and sub-typing of MAP field isolates. The methodology MALDI-TOF (Matrix-assisted laser desorption/ionization coupled with Time-of-Flight mass spectrometer), applied to the micro-organism "in toto" seems to be able to identify all known bacteria. This method has been successfully applied for the identification of many species among the genus Mycobacterium[3]. The objective of this work was to explore the use of MALDI-TOF for rapid identification and sub-typing of MAP isolates.In this preliminary study, we considered a reference strain (ATCC 19698) and 8 MAP field isolates, from different source and hosts. All isolates were type II. For MALDI-TOF analysis, we used a previously developed protocol [3]. Samples were submitted to MALDI-TOF (Bruker Microflex LT, Germany) and the spectra, for each strain, were assessed considering the results of the 5 technical replicates. The obtained spectra were standardized and aligned using the program Biotyper 3.0 (Bruker) for the identification of the species. According to a customized database containing spectra from different mycobacteria, a good identification score was obtained. We further evaluated whether this approach could be useful for MAP sub-typing, but our data suggested a random distribution of the spectra obtained from the individual replicate with respect to different strains, not allowing to differentiate any cluster within the strains. In future we are planning to increase the number of MAP (including Type I strains) and other mycobacteria strains analysed in order to imporove our database. Founded by Italian Ministry of Health, PRC2013/017 (E52I14000780001).1. Castellanos E et al. Progress in molecular typing of Mycobacterium avium subspecies paratuberculosis. Res Vet Sci 2012, 92(2):169-179.2. Ahlstrom C et al. Limitations of variable number of tandem repeat typing identified through whole genome sequencing of Mycobacterium avium subsp. paratuberculosis on a national and herd level. BMC Genomics 2015, 16:161.3. Shitikov E et al. Mass spectrometry based methods for the discrimination and typing of mycobacteria. Infect Genet Evol 2012, 12(4):838-845.

Keywords : MALDI-TOF, identification

P-03.28
Performance of serological assay using specific proteins of Mycobacterium avium subspecies paratuberculosis for the diagnosis of Johne's disease in large ruminants

Sujata Jayaraman[1], K K Chaubey[2], Rinkoo D Gupta[3], Manju Singh[1,2], Mukta Jain[1], B J Stephen[2], Saurabh Gupta[2], S V Singh[2], G K Aseri[1], Manali Datta[4], Neelam Jain[4], Parul Yadav[1], Neeraj Khare[1], **J S Sohal**[1]

[1]Amity Institute of Microbial Technology, Amity University Rajasthan, Jaipur, Rajasthan, India. [2]Animal Health Division, Central Institute for Research on Goats, Makhdoom, PO – Farah, Dist. – Mathura, Uttar Pradesh, India. [3]South Asian University, New Delhi, India. [4]Amity Institute of Biotechnology, Amity University Rajasthan, Jaipur, Rajasthan, India

Paratuberculosis caused by Mycobacterium avium subspecies paratuberculosis (MAP) is a major production infection of ruminants. Control practices are largely dependent on 'Test and Cull' method but due to lack of efficient diagnostics, this policy has not yielded results to the theoretical potential. Therefore, diagnostic tests require improvements in terms of sensitivity and specificity. Serology based assays are cheapest and can be adopted as 'Herd Screening Test' using ELISA based system. With this aim four proteins in the MAP genome (coded names; JSK-I, JSK-II, JSK-III & JSK-IV) were shortlisted as potential diagnostic markers. Gene sequences for these proteins were optimized for expression in E. coli system. Bulk production of these proteins was done in E. coli system. All these proteins were identified as strong antigens using immune-proteomic analysis with previously characterized serum samples from cattle, buffaloes and camels. These proteins were used to optimize ELISA based test for the diagnosis of paratuberculosis in large ruminates. The new ELISA test using these proteins was extended to clinical samples from different geographical regions of country. The test was found to be high sensitivity and specificity as compared to conventional 'Indigenous ELISA' utilizing semi-purified protoplasmic antigens (sPPA) and also had high correlation with histo-pathological lesions.

Keywords :
Paratuberculosis, large ruminants, diagnosis, ELISA, specific antigens

P-03.29
A new serological test for the diagnosis of paratuberculosis infection in small ruminants using specific recombinant antigens

Sujata Jayaraman[1], K K Chaubey[2], Rinkoo D Gupta[3], Manju Singh[1,2], Mukta Jain[1], B J Stephen[2], Saurabh Gupta[2], S V Singh[2], G K Aseri[1], Manali Datta[4], Neelam Jain[4], Parul Yadav[1], Neeraj Khare[1], **J S Sohal**[1]

[1]Amity Institute of Microbial Technology, Amity University Rajasthan, Jaipur, Rajasthan, India. [2]Animal Health Division, Central Institute for Research on Goats, Makhdoom, PO – Farah, Dist. – Mathura, Uttar Pradesh, India. [3]South Asian University, New Delhi, India. [4]Amity Institute of Biotechnology, Amity University Rajasthan, Jaipur, Rajasthan, India

Small ruminants (goat and sheep) are important animals for the livelihood and nutritional security of large number of marginal and landless farmers of the country. Pathogens like Mycobacterium avium subspecies paratuberculosis (MAP) are primarily responsible for low per animals productivity of our native livestock breeds. A recent study indicated that due to MAP there is loss of Rs 1,840 (US$ 38.33) per infected sheep per year. Presently used control practices (Test & Cull) are not sufficient due to control disease due to vertical transmission of infection and lack of efficient diagnostics. Therefore, improved cheap diagnostics are the need of hour. We shortlisted four proteins in MAP genome (names coded as JSK-I, JSK-II, JSK-III & JSK-IV) as potential diagnostic markers that are absent from genome of M. bovis. Bulk production of these proteins was done in E. coli system. All these proteins were identified as strong antigens using immune-proteomic analysis with previously characterized serum samples of sheep and goats. All these proteins were used to optimize ELISA based test to diagnose paratuberculosis infection in small ruminants. ELISA test using these proteins were extended to clinical samples from different geographical regions of country. Results showed that the sensitivity and specificity of this new ELISA test was highly improved as compared to conventional indigenous ELISA test using semi-purified protoplasmic antigen sPPA) and had high correlation with histo-pathological lesions.

Keywords :
Paratuberculosis, small ruminants, diagnosis, ELISA, specific antigens

P-03.30

Identification of Mycobacterium avium subspecies paratuberculosis in milk products (Paneer or Fresh cheese) from local dairies in Farah, Mathura and Agra employing six diagnostic tests

Bjorn John Stephan[1], Manju Singh[1], Saurabh Gupta[1], Kundan Kumar Chaubey[1], Sahzad[1], **Shoor Vir Singh**[1], Sujata Jayaraman[2], Mukta Jain[2], Gajendra Kumar Aseri[2], Manali Datta[2], Neelam Jain[2], Parul Yadav[2], Neeraj Khare[2], Kuldeep Dhama[3], Jagdip Singh Sohal[2]

[1]Animal Health Division, Central Institute for Research on Goats, Mathura, Uttar Pradesh, India. [2]AIMT & AIB, Amity University Rajasthan, Jaipur, India. [3]Division of Pathology, Indian Veterinary Research Institute, Izatnagar, Uttar Pradesh, India

Johne's disease is chronic progressive enteritis, caused by Mycobacterium avium subspecies paratuberculosis (MAP) in domestic ruminants world-wide. Persistence live MAP bacilli in pasteurized milk and milk products has been reported regularly in global journals. Therefore MAP is the cause of major concern for public health and food safety. Regular consumption of contaminated milk and milk products could lead to Crohn's Disease, a systemic enteritis associated with MAP infection in human beings. In this study, 55 paneer (fresh cheese) samples collected from local dairies in Farah, Mathura and Agra region were screened by six tests (traditional and newly developed and standardized) for the presence of MAP bacilli. Of the 55 samples screened, 54.5 (30), 32.7 (18), 54.5 (30), 74.5 (41), 50.9 (28) and 63.6% (35) samples were positive by acid fast staining, IS900 PCR, indigenous plate ELISA, indigenous dot-ELISA, Latex agglutination test and Fluorescent antibody test, respectively. Results revealed that the paneer sold by the local dairies had high bio-load of MAP. This was due to the use of milk for the preparation of paneer by the diaries was contaminated with MAP bacilli in view of the high bio-load of MAP in domestic livestock. Therefore it is imperative to first control MAP infection in domestic livestock so that milk produced by animals is free of bacilli. Otherwise milk will continue to be infected with MAP bacilli and use of such milk will not be safe for human consumption. Isolation of the infected animals is not a practical solution in India, since majority of animals are infected with MAP in the absence of National control programs. Therefore we strongly recommend that during the preparation of paneer, the milk should be boiled properly to inactivate the MAP bacilli to prevent it from entering the human food chain.

Keywords :
Paratuberculosis, cheese, IS900 PCR, Indigenous plate ELISA, Indigenous dot-ELISA, Latex agglutination test, Fluorescent antibody test

P-03.31

Use of the IDvet complete PCR testing solution to manage Map infection in bovine herds

Lise Grewis, Pourquier Philippe[1], Laffont Mathieu[2]

[1]IDvet. [2]IDvet Genetics, France

Objectives :IDvet Genetics has developed an easy-to-use Map detection system, including a rapid sample preparation kit (EZ-Prep), and a qPCR test, ID Gene TM Paratuberculosis Duplex (IDMAP). The IDMAP test may be used for quantitative or relative quantitative testing, for individual or pooled testing. The IDMAP PCR may be used as a decision-making tool for managing positive animals. The PCR test has a limit of quantification of the complete method of 1000 mycobacteria per gram of faeces. The objective of this study was to demonstrate how this complete PCR system may be used in the field. Materials and Methods:Faeces sample preparation was performed using the EZ-PREP system on 5g of sample.Two suckling herds were tested. The first herd (n=161) had a sero-prevalence of 10%, and the second (n=40) of 3%. All animals were tested individually and pooled by PCR, and individually using the (ID Screen®) Paratuberculosis Indirect ELISA as per manufacturer's instructions. Results :For herd 1 (seroprevalence of 10%), excellent agreement between serology and PCR for strong positive animals (high S/P% and early Cq results) was obtained. These animals were heavy shedders which represent the highest risk for spreading disease in the herd. Quantitative testing allows farmers and veterinarians to identify heavy and chronic shedders which are sero-negative. The use of pooling identifies chronic and intermittent shedders and passive carriers which would not be detected by serological testing. Results for herd 2 with a seroprevalence of 3% demonstrate that animals may be detected positive by PCR and negative by serology. These animals were considered to be intermittent shedders with little risk of spreading the disease to other animals. The decision to cull the animal will be made by the farmer depending on the need to lower the Map pressure in the herd. Conclusion :Quantitative testing allows for decision-making based on the detection of high shedders and/or chronically-infected animals with sero-negative results. The use of pooled samples allows for the detection of intermittent shedders or passive carriers. The method may be used in different context depending on the herd status and aim of the control program.

Keywords :
PCR, paratuberculosis, managing

NOTES

NOTES

NOTES

NOTES

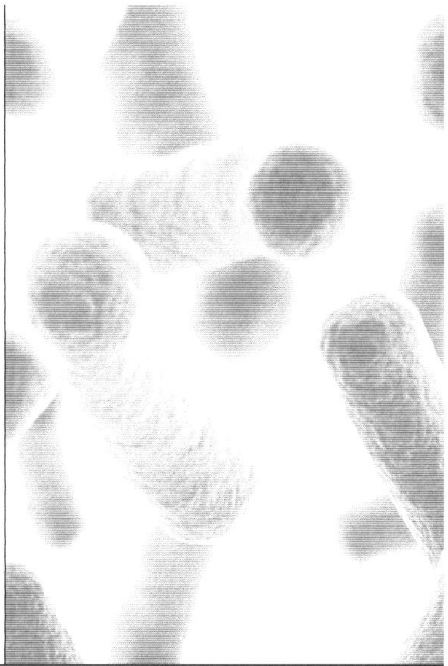

SESSION 4

Exposure & Transmission

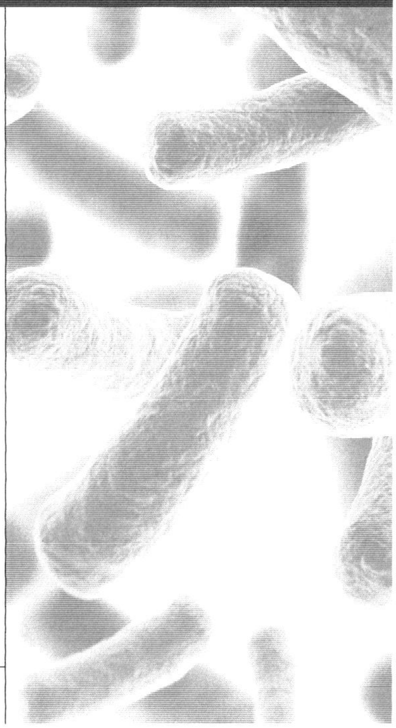

TRANSMISSION DYNAMICS OF MAP IN PASTORAL GRAZING SYSTEMS OF RUMINANT LIVESTOCK IN NEW ZEALAND

Cord Heuer, Massey University, New Zealand

Introduction

Clinical Paratuberculosis (PTB) is caused by extensive multiplication of Mycobacterium avium subspecies paratuberculosis (MAP) in the intestinal wall of the distal jejunum and ileum eliciting a hyperimmune reaction that results in gut dysfunction, nutrient malabsorbtion, diarrhoea, chronic wasting and death. However, the clinical form of PTB is rare. By far the most of all MAP infections are tolerated by the host with either small temporary ('sub-clinical') production loss or no health impact at all. This severity pattern appears to apply to PTB of all domestic ruminant production systems regardless of eco-climatic region or intensity of production. Thus, the infection prevalence and clinical incidence observed in New Zealand (NZ) may reflect a worldwide pattern (**Table 1**).

Table 1 : Average herd or flock prevalence of MAP infection and annual clinical PTB incidence in four New Zealand production systems (95% confidence interval CI)

Farm type	Herd/flock prevalence		Clinical incidence[1]		Source
	Mean	95% CI	Mean	95% CI	
Dairy cattle	54%	44-68%	0.50%	0-6.2%[2]	Hunnam 2014
Beef cattle	42%	35-50%	0.20%	0.1-.3%	Verdugo et al. 2014a
Sheep	76%	70-81%	0.20%	0.1-0.2%	Verdugo et al. 2014a
Deer	46%	38-55%	0.40%	0.2-0.7%	Verdugo et al. 2014a

[1] *Infected herds/flocks (cases/100 breeding females+year)* [2] *range*

Transmission patterns of all-year pasture based systems with strictly seasonal calving/lambing are likely to be very different from those in non-seasonal, mostly indoor, concentrate based intensive systems. Pasture based systems are typical for Australia and almost exclusively found in New Zealand and Chile with mild winter climates allowing for extended MAP survival. Intensive systems are predominant in most Northern hemisphere countries with below zero Celsius temperatures for several months. These two most common systems are associated with different exposure patterns to MAP. Pastoral dairy farming for example mainly exposes susceptible calves to their dams for 2-3 days at calving, to transiently shedding calves in collective calf pens to 3 months of age and to contaminated pasture by MAP shedding cows or wildlife species (Nugent et al. 2011) up to 6 months of age. In intensive dairy systems on the other hand, calves are removed from dams within a few hours after calving, then often housed individually without exposure before being transferred to collective calf pens for about 2 months with high contact opportunity to transiently shedding pen mates, but usually little exposure thereafter (**Table 2**).

Pastoral beef, sheep and deer farming is different from pastoral dairy management in that susceptible offspring remain with their dams for 3 (sheep) or 6 months (beef, deer). Moreover, sheep and beef cattle are often set stocked for several months on the same pasture from calving/lambing up to weaning with ample opportunity for dam-offspring transmission. However, despite this extended and close dam-calf/ewe-lamb contact, the incidence of clinical PTB in beef cattle and sheep was relatively low compared to other species (**Table 1**). This raised the question whether MAP type I (or S) may be less virulent and whether it may be transmitted from sheep to cattle. A study to explore this transmission hypothesis is described in the following section.

Table 2 : Transmission patterns of dairy cattle in non-seasonal calving, concentrate based systems vs. seasonal calving, pasture based systems

Source of transmission	Concentrate based systems	Pasture based systems
Vertical	P ~ 0.04 to 0.15	P~ 0.04 to 0.15
Pseudo-vertical	Low as calf removed from dam almost immediately P ~ 0	High as calf removed from dam after 2-3 days P ~ 0.1/0.4 for low/high shedders
Transient calf-calf	Age 0- 1/2 months in individual pens 1/2 - 2.5 months in collective calf pens; 2m HIGH EXPOSURE	Age 3d – 3 months in collective calf pens; 3m HIGH EXPOSURE
Weaning	~10 weeks	~13 weeks
Environmental	Age >2.5m penned with older (non-shedding) young stock (<24m); 10m LOW EXPOSURE Limited exposure to pasture	Age 3 – 6m on pasture contaminated by adult cows; 3m HIGH EXPOSURE Age 6-24m on 'run-off' without exposure to adult pasture; exposure to MAP infected wildlife? (Nugent et al. 2011)

Transmission between species through co-grazing

Strain differentiation of MAP by molecular methods has provided the tool to explore transmission pathways. A variety of nucleotide sequence based techniques were applied to differentiate MAP strains. The methods were reviewed by Stevenson 2015. They include variable number of tandem repeats (VNTR), short sequence repeats (SSR), multilocus short-sequence-repeats (MLSSR), mycobacterial interspersed repetitive unit (MIRU), whole genome sequencing (WGS) and others (Ahlstrom et al. 2011; Mitchell et al. 2016).

Using VNTR/SSR (Collins et al. 2012) for typing isolates from pooled faecal samples of healthy adult animals of randomly selected herds/flocks, Verdugo et al. (2014b) showed that about 80% infected beef herds were infected with ovine (S) MAP type I, typically found in sheep. Proportional similarity index analysis of these data revealed that MAP strains from beef cattle and sheep were similar providing strong evidence for cross-species transmission from sheep to cattle while grazing the same pasture.

Figure 1 : Proportional similarity index evaluating similarity (>0.5) and dissimilarity (<0.5) of VNTR/SSR strains of MAP from pooled faecal samples of four paired host species comparisons (error bars are 95% confidence intervals).

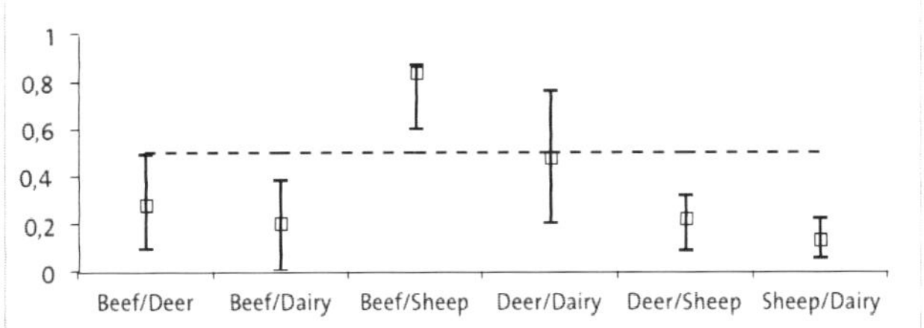

Conversely, a dissimilarity of strains of most other species comparisons indicated that transmission was less likely between those hosts (**Figure 1**). However, the dissimilarity of beef/deer strains changed to similarity, albeit not significantly, when the analysis was restricted to farms where the two species were grazed together or in succession.

Transmission between farms through animal movement

Encouraged by these results, a movement network was developed from data of a large corporate NZ farm enterprise (Landcorp Ltd.) operating 112 dairy, beef, sheep, and/or deer properties. Faecal culture isolates of MAP from 33 farms were strain typed by VNTR/SSR and the probability of farms sharing the same strain regressed on the proximity in the network. The two most frequent strains are displayed geographically (**Figure 2, left**) and in the network (**Figure 2**, centre). After adjusting for species, island (N/S), spatial distance, and number of animals moved, the path length was negatively correlated with the probability of two connected farms sharing the same strain (**Figure 2, right**). Thus, the closer two farms were connected in the network the more likely they shared the same strain. This was strong evidence that MAP was transmitted between farms despite a relatively high level of endemic prevalence of MAP infected farms (**Table 1**).

Figure 2: Locations (left) of two dominant MAP strains in a movement network of n=33 properties (centre) isolated from N=112 farms of one corporate NZ farmer (Landcorp Ltd.). The proximity of farms (path length) was negatively associated with the probability of sharing the same strain (right, red dots are means, blue dots 95%CI).

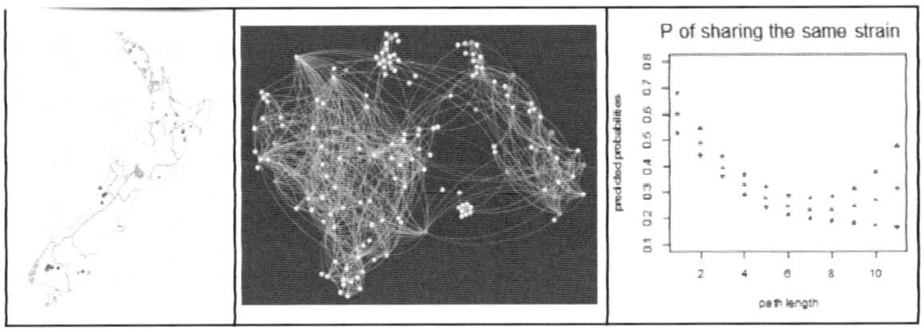

We conclude that strain typing, movement networks and appropriate statistical analysis are powerful tools for evaluating the transmission of MAP between animals, species, and farms. The transmission of potentially less virulent MAP strains from sheep to beef cattle could be relevant for interventions to control PTB by co-grazing if it can be shown that MAP type I (or S) was less virulent for cattle. As animal movements increase the risk of MAP transmission between farms, selecting animals from farms with a low infection risk can decrease the endemic level of PTB.

Acknowledgements

This research was funded by the NZ Johne's Research Consortium (JDRC) which was ably managed by Kaylene Larking. I like to point out the contribution of several individuals, especially PhD students Cristobal Verdugo, Nelly Marquetoux and Milan Gautam, my colleagues Des Collins, Geoff De Lisle and Marian Price-Carter of AgResearch LtD., Eve Pleydell and Nigel French at the mEpiLab/Hopkirk Institute of Massey University, Peter Wilson and Anne Ridler at Massey's Institute of Veterinary, Animal and Biomedical Sciences, and Gordon Williams of Landcorp Ltd. for the provision of movement data spanning four years.

References

1. *Ahlstrom C, Barkema HW, Stevenson K, Zadoks RN, Biek R, Kao R, Trewby H, Haupstein D, Kelton DF, Fecteau G, Labrecque O, Keefe GP, McKenna SLB, Buck JD. Limitations of variable number of tandem repeat typing identified through whole genome sequencing of Mycobacterium avium subsp. paratuberculosis on a national and herd level. BMC Genomics 16:161, 2015*
2. *Collins DM, Price-Carter M, Scandurra GM, DeLisle GW. DNA typing of recent New Zealand isolated of Mycobacterium avium subsp. paratuberculosis from cattle, sheep and deear by VNTR and SSR. Proceedings of the 11th International Colloquium on Paratuberculosis, Sydney, Australia, 5-10 February 2012: 251-252, 2012*
3. *Hunnam J. Johne's control is possible: Does anyone have a MAP? In: NZVA Annual Conference. Hamilton. Pp 133-50. VetLearn Foundation, 2014*
4. *Nugent G, Whitford EJ, Hunnam JC, Wilson PR, Cross ML, de Lisle GW. Mycobacterium avium subsp. paratuberculosis infection in wildlife on three deer farms with a history of Johne's disease. New Zealand Veterinary Journal 59(6), 293–298, 2011*
5. *Stevenson K. Genetic diversity of Mycobacterium avium subspecies paratuberculosis and the influence of strain type on infection and pathogenesis: a review. Veterinary Research 46:64, 2015*
6. *Verdugo C. Epidemiology of Mycobacterium avium subspecies paratuberculosis infection on sheep, beef cattle and deer farms in New Zealand. PhD thesis, Massey University, Palmerston North, New Zealand, 2013*
7. *Verdugo C, Jones G, Johnson WO, Wilson PR, Stringer L, Heuer C. Estimation of flock/herd-level true Mycobacterium avium subspecies paratuberculosis prevalence on sheep, beef cattle and deer farms in New Zealand using a novel Bayesian model. Preventive Veterinary Medicine 117, 447-55, 2014a*
8. *Verdugo C, Pleydell E, Price-Carter M, Prattley D, Collins D, de Lisle GW, Vogue H, Wilson P, Heuer C. Molecular epidemiology of Mycobacterium avium subsp paratuberculosis isolated from sheep, cattle and deer on New Zealand pastoral farms. Preventive Veterinary Medicine 117, 436-46, 2014b*

ORAL

O-04.1
Transmission of infection among group-housed calves inoculated with Mycobacterium avium subsp. paratuberculosis

Caroline Corbett[1], Jeroen De Buck[1], Karin Orsel[1],Herman Barkema[1]

[1]*University of Calgary, Faculty of Veterinary Medicine, Department of Production Animal Health. 3330 Hospital Drive NW, T2N 4N1, Calgary, Alberta, Canada*

Control programs for Johne's disease are primarily based on preventing transmission from cow to calf to reduce new infections on farm; however, the potential transmission of Mycobacterium avium subspecies paratuberculosis (MAP) among the calves that are penned together is largely overlooked. An experimental transmission trial was carried out with the following goals: 1) to quantify the extent to which transmission of MAP occurs to penmates among group-housed calves; and 2) to determine the onset and duration of MAP shedding after exposure to inoculated calves.
32 newborn Holstein-Friesian bull calves were allocated into pens of 4. In 7 of these pens, 2 calves were inoculated with 5x108 CFU's of MAP (IN), and 2 calves acted as contact-exposed (CE). The 8th pen of 4 was designated as a non-infected control group. Calves were group housed for 3 months to allow for any potential fecal oral transmission to occur, during which time fecal samples were collected 3 times a week, and blood and environmental samples weekly. After 3 months, the CE calves were individually penned for 3 more months until the end of the trial to ensure any new infections would be detectable. During the last 3 months, fecal and blood samples were collected weekly. Tissue samples were collected at necropsy. Both fecal and tissue samples were cultured, followed by MAP-specific qPCR.
Following group housing with IN calves, all CE calves began to shed MAP, detected by positive fecal culture. Fecal shedding of the CE calves ceased after individual housing; however, 7 of 14 CE calves sampled at necropsy 3 months later after individual housing were MAP tissue culture-positive. 2/14 calves had more than one tissue sample positive. These results indicate that CE calves can become infected following 3 months of exposure to IN penmates.
The results obtained from this study generate new knowledge regarding patterns of shedding and tissue infection in either CE or IN penmates and may lead to more effective management practices, minimizing the spread of JD.

Keywords :
Johne's disease, transmission, calf, calves, group-housing, transmission trial

O-04.2
Prevalence of Mycobacterium avium subsp. paratuberculosis (MAP) in pasteurised milk and evidence for mechanisms of survival

Saeedi A.[1], Gerrard Z.[1], Swift B.[2], and Rees C.[1]

[1]*University of Nottingham, School of Biosciences, Sutton Bonington Campus, Leics, LE12 5RD.* [2]*University of Nottingham, School of Veterinary Science and Medicine, Sutton Bonington Campus, Leics, LE12 5RD, UK*

The association between MAP and Crohn's disease means that pasteurised milk has been highlighted as a key vector for entry of MAP into the human food chain since there is good evidence that the organism is able to survive commercial pasteurisation. Here we report a survey of pasteurised semi-skimmed milk (1.7 % fat) using the phage-PCR method. Samples (385) were purchased from local retail outlets and viable mycobacteria was detected in 31% using the phage-PCR method (Botsaris et al., 2013). A third of these were confirmed to contain MAP by plaque-PCR (approximately 10% of all samples) but only 1% of samples had more than 10 cells per 50 ml.
To investigate the mechanism of survival, the distribution of MAP cells in naturally contaminated milk was investigated. Levels of MAP in replicate samples were compared with or without lysis of somatic cells. For 10 samples where no lysis was carried out before testing, a very low correlation was seen between the results from paired samples. In contrast, when somatic cells were lysed prior to sampling the results became more consistent (r2 =0.78). This result indicates that prior to somatic cell lysis, the MAP cells are not evenly distributed in the milk and become more evenly dispersed in the sample after release from the somatic cells, consistent with internalisation of MAP.
The phage assay was also used to rapidly monitor heat inactivation of M. smegmatis in spiked milk at 72 °C. Interestingly there was not a complete correlation with viable count and the phage assay results. The viable count data indicated that complete inactivation occurred, but the phage assay was able to detect low numbers of viable cells which survived heating up to 30 s. Sub-lethally injured cells may be able to support phage growth, and therefore will be detected by the phage assay. However they may not be able to form colonies on solid media and this result may help to explain some of the differences gained between laboratory-based determination of heat resistance and the consistent reports of MAP survival of this process and its detection in retail milk.

Keywords :
Pasteurisation, pasteurised milk, survival, Crohn's disease, Phage-PCR assay, sub-lethal injury

O-04.3

Regional Differences in Johne's Disease Prevalence based on Environmental Culture and Bulk Milk Testing

D. Kelton[1], H. Barkema[2], C. Bauman[1], C. Pickel[2]

[1]University of Guelph, Guelph, Ontario, Canada. [2]University of Calgary, Calgary, Alberta, Canada

The first Canadian National Dairy Study (NDS) was completed in 2015 with an overarching objective to benchmark the health, productivity and management of the national dairy herd. The study included over 1,340 dairy farms (11% of all dairy farms in Canada), of which 46% had participated in a voluntary regional Johne's disease control program. Regional programs in Canada are based on either fecal culture/PCR of environmental samples, or cow/bulk tank milk ELISA tests, limiting the ability to compare herd-level prevalence estimates among regions. As part of the NDS, 375 farms in all 10 provinces were visited and both environmental fecal and bulk tank milk samples were collected for testing.
A composite manure sample was collected from three different areas on each farm: the breeding age heifer pen, the milking cow area (alleyways), and from the manure storage area (either the liquid manure pit or the manure pile). Each of these three samples underwent DNA-based (PCR) testing for the presence of Mycobacterium avium ssp. paratuberculosis. Bulk tank milk samples were collected either at the time of the herd visit by study personnel, or after the visit by accessing payment samples collected by milk transporters. Milk was tested with a commercial ELISA kit.
Based on results of environmental fecal testing, the prevalence of test-positive farms was highest in Western Canada and Ontario (20%), moderate in Eastern Canada (12%) and lowest in Quebec (5%). Results of bulk tank milk testing are pending and will be presented for comparison. Recognizing that these herd level tests lack sensitivity and are likely only detecting herds with a higher within-herd prevalence of infection, the regional differences are of interest and may be related to herd size and/or housing systems. Data available from the NDS will be used to identify factors associated with these differences.

Keywords :
Prevalence, milk testing

O-04.4

True within herd prevalence distribution and risk factors to MAP infection among dairy herds in Chile

Cristobal Verdugo[1], Maria Francisca Valdez[1], Miguel Salgado[1]

[1]Instituto de Medicina Preventiva Veterinaria, Universidad Austral de Chile, Chile

The estimation of MAP true prevalence and the determination of risk factor for infection are two basic elements to develop a sound control program. Spite of the infection is endemic in the Chilean Dairy sector, little is known about MAP prevalence, and risk factors for infection have not been determined for this disease in Chile. Therefore, the objective of this study was to estimate the within herd true prevalence (TP) distribution among MAP infected farms, and determine risk factors associated to MAP infection in the Chilean Dairy sector.
All lactating cows on 42 know infected farms were serum-ELISA tested. The selected farms, were randomly selected from a list of 300 infected farms, from a previous study. A Bayesian latent class model was used to estimate TP distribution. Similarly, a Bayesian logistic regression model was used to determine risk factors associated to infection. Posterior Bayesian probabilities were used to assess statistical important association between risk factors and the probability of MAP infection. The two models adjust ELISA test results by its sensitivity (Se) and specificity (Sp) performance, obtaining unbiased distributions of infection prevalence and risk factors. Priors Se, Sp, & TP distributions were elicited by a combination from literature research and expert opinion.
A total of 5,494 animals were ELISA tested, where 348 were ELISA positive. The posterior distribution indicated an overall median TP prevalence of 9%, and a 95% posterior probability distribution of 3% - 25%. The Bayesian regression model identified two risk factors associated to MAP infection. The use of pool of milk to feed calves increases the risk by 3 times. Whereas, the proximity of calves pen to the milking parlor was positively associated, where the risk of infection increased 1.1 times per each meter of distance decrease.

Keywords :
Prevalence, Risk Factors, Dairy

POSTER

P-04.1
MOLECULAR ANALYSIS OF MYCOBACTERIUM AVIUM SUBSP. PARATUBERCULOSIS ISOLATES FROM KOREAN CATTLE FARMS USING MIRU-VNTR TYPING METHOD

Hong-Tae Park[1], Seung Won Shin[1], Myunghwan Jung[1], Hyun-Eui Park[1], Yong-Il Cho[2], **Han Sang Yoo**[1]

[1]*Department of Infectious Diseases, College of Veterinary Medicine, Seoul National University, Seoul, 08826, Korea.* [2]*National Institute of Animal Science, Rural Development Administration, Cheonan, 31000, Korea*

Johne's disease, also called paratuberculosis (PTB) is caused by Mycobacterium avium subsp. paratuberculosis (MAP) which is economically important disease in cattle. Understanding a genetic diversity of MAP isolates is important to the epidemiological studies because it can reveal the infection sources and transmission route and therefore may contribute to the disease control. The aim of this study was to investigate the genetic diversity of MAP isolated from Korean cattle farms using PCR and restriction based methods.Total of 18 MAP strains from six farms in three regions were isolated using VersaTREK ParaJEM system. Positive resulted samples were confirmed by IS900 & ISMAP02 PCR then sub-cultured in the modified Herrold's egg yolk medium. Colonies were subjected to downstream analysis, the IS1311 PCR-restriction enzyme analysis (REA) and mycobacterial interspersed repetitive unitsffvariable number tandem repeat analysis (MIRU-VNTR). The MIRU-VNTR typing was performed using eight polymorphic loci (292, X3, 25, 47, 3, 7, 10 and 32).From the IS1311 PCR-REA, two MAP strains were identified as "cattle type" and 16 strains were "bison type". In the MIRU-VNTR, all of the cattle type strains were identified as INMV 2 (32332228) and all of the bison type strains were identified as INMV 68 (22532228). The results revealed that cattle type strains and bison types were coexisted in one farm that twelve isolates were obtained. The six isolates from rest of five farms were all identified as bison type. Although the number of isolates was low, it is supposed to be a major types of MAP are bison type in Korea. Furthermore, previous studies reported that the bison type of MAP were found in Korean black goats and wild boar.In the present study, genetic diversity of MAP isolates obtained from Korean cattle farms were analyzed by two different molecular typing methods. From the results, the isolates were discriminated as only two types. Therefore, the new molecular typing methods which have high discrimination index should be applied in further study.This work was carried out with the support of Cooperative Research Program for Agriculture Science & Technology Development (PJ008970) RDA, BK21 PLUS Program and Research Institute for Veterinary Science, SNU, Republic of Korea.

Keywords : MAP, Molecular analysis, MIRU-VNTR

P-04.02
ACCOUNTING FOR INDIVIDUAL CHARACTERISTICS IN MODELLING THE WITHIN-HERD SPREAD OF BOVINE PARATUBERCULOSIS

Guillaume Camanes[1,2], Alain Joly[2], Racem Ben Romdhane[1], Gaël Beaunnee[1,3], Pauline Ezanno[1]

[1]*INRA, ONIRIS, UMR1300 BioEpAR, CS40706, F-44307 Nantes, France.* [2]*Groupement de Défense Sanitaire de Bretagne, 56019 Vannes, France.* [3]*INRA, UR1404 MaIAGE, F-78352 Jouy-en-Josas, France*

Controlling the spread of Mycobacterium avium subsp. paratuberculosis (Map) within and betweenherds still represents a major economic challenge. However, the succession of infection stages and the individual shedding pattern are not well known for infected animals, whereas this precise knowledge is required to identify effective control measures in order to limit new infections.Combining field observations and epidemiological modelling appears to be a relevant approach toproduce such knowledge. However, most existing models do not account for individual characteristics of animals (health status, shedding level), or neglect shedding by infected calves and Map survival in the environment, the latter also being a pivotal process involved in Map persistenceat the herd scale. We propose a new individual-based model that accounts for all of the available knowledge on Maptransmission and shedding. By considering precisely individual characteristics of animals and their variation over time, this model can be calibrated on observed longitudinal field data. Our individual-based model is a stochastic model in discrete time, with a time step of one week. It has been programmed in C++ to ensure computing efficiency. It couples population and infection dynamics within a dairy cattle herd, and considers explicitly all of the farming environments and Mapsurvival. It accounts for the five routes of transmission (in utero, contaminated colostrum/milkingestion, and ingestion of contaminated faeces present in the local and in the general environments), as well as for the early shedding of Map by infected calves. Animals are individually characterized, with information about their age group (unweaned calves, weaned calves, youngheifers, heifers, and adult cows), their progression between health statuses (susceptible, resistant, transiently infectious, latently infected, subclinically infectious, and clinically affected), their contact with farming environments, their shedding level, and their detection test history. With our individual-based model, we aim at assessing biological assumptions, notably concerning the probability of adult infection, shedding dynamics, and the succession between infection statuses.

Keywords : P aratuberculosis, model, individual-based, Map, spread, transmission

P-04.3

Use of liquid culture with IS-900 PCR confirmation on fecal samples to assess exposure of young calves to Mycobacterium avium subsp. paratuberculosis in infected herds

Emmanuelle de Marchin[1], Grégoire Fabien[1], Evrard Julien[1], Houtain Jean-Yves[1]

[1]*ARSIA (Regional Association for Animal Identification and Health), Ciney, Belgium*

Introduction : Johne_disease elimination from infected herds takes several years and requires to combine a « test and cull » strategy in adult cows with application of biosecurity measures in young calves in order to prevent new infections. The acceptance of these measures by breeders is the main difficulty in the field since it often has a significant impact on management habits. The objective of this study was to investigate in field conditions the use of liquid culture with IS-900 PCR confirmation on fecal samples to assess exposure of young calves to Mycobacterium avium subsp. paratuberculosis (MAP) from infected herds.

Material and methods : In 10 herds involved in the South-Belgian regional Johne_disease control plan and containing at least 10 % of MAP shedders among adult cows, 141 calves were sampled at days 7, 30 and 60 of life. At each sampling, data about the origin of colostrum or milk given, the rearing method (suckling vs non-suckling) applied, as well as the housing-type of the calf, were collected. Each fecal sample was cultured using para-JEM® liquid culture system for 42 days. Presence of MAP DNA was assessed using an IS-900 real-time PCR. Univariate logistic regression was used to assess risk factors associated with calves exposure to MAP.

Results : During this study, MAP could be detected in at least one fecal sample of 24 calves (17%; IC95%: 10.7-23.4). 2 calves were detected positive twice (days 7 and 30). The rearing method was the only risk factor associated with exposure to MAP in this study: suckling calves were 5.9 (IC95%: 2.3 - 15.2) more at risk to be exposed to MAP during their first two months of life than non-suckling calves. No significant effect could be observed for the proportion of shedders among adult cows in the herd, neither for the type of colostrum or milk (maternal vs commercial) feeding or the type of housing.

Conclusions : This study showed that liquid culture with IS-900 PCR confirmation is sensitive enough to detect MAP in faeces of young calves in infected herds which opens new perspectives in management of herd infected by MAP.

Keywords : PCR, para-JEM®, calves, management, exposure

P-04.4

Which phenotypic traits of dairy cattle resistance to bovine paratuberculosis can enhance disease control at herd scale ?

Racem Ben Romdhane[1], Guillaume Camanes[1], Raphaël Guatteo[1], Gaël Beaunee[1,2], Christine Fourichon[1], Pauline Ezanno[1]

[1]*INRA, Oniris, UMR1300 BioEpAR, CS40706, F-44307 Nantes, France.* [2]*INRA, UR1404 MaIAGE, F-78352 Jouy-en-Josas, France*

Control strategies of bovine paratuberculosis at herd scale are currently based on hygiene improvement and test-and-cull of infected animals. These actions are difficult to implement by farmers and lack effectiveness to reduce infection prevalence. Genetic selection could enhance disease control if focused on phenotypic traits of animals which influence the disease dynamics in a population. However, there is a lack of knowledge on which traits contribute the most to the disease dynamics.Our objective was to assess which phenotypic traits should be targeted to enhance paratuberculosis control in dairy cattle herds.We used a within-herd transmission model where we modified traits related to the course of infection dynamics in an animal. The model is stochastic and simulates herd population dynamics and infection course in animals. It accounts for all known transmission routes, shedding heterogeneity, and age related susceptibility. We studied model outcomes describing the disease dynamics in the herd: infection persistence, prevalence, proportion of affected animals, and cumulated number of newly infected animals, 25 years after infection first introduction. We compared a situation assuming phenotypic traits at their current values with situations where animals more resistant to infection would have been successfully selected. Each scenario was defined as a combination of improved phenotypic traits (i.e. reduced susceptibility and reduced duration of the susceptible state, shedding level and durations in infection stages). Cluster analyses were performed to highlight which traits must be improved and to what extent on, to contain the disease dynamics in the herd.Phenotypes combining a high reduction of susceptibility (e.g. halved) with short susceptibility duration (e.g. halved) were the most effective, lowering the probability of infection persistence in the herd and the prevalence in infected herds (e.g. decreased of 60% and 75%, respectively). If, in addition, the duration before entering the affected stage was increased (e.g. doubled) and the shedding level of affected animals was decreased (e.g. halved), the probability of persistence and the prevalence further decreased (e.g. by 92% and 99%, respectively). Modifying other traits was much less effective.Gaining knowledge on phenotypic traits of resistance to be targeted is useful to orientate future efforts in genetics. Regarding the long delays of genetic selection, we should evaluate the potential effectiveness and time needed to control paratuberculosis when combining it with current control measures, starting from an enzootic situation. This study was funded by INRA (métaprogramme GISA), Apis-Gene, and GDS France.

Keywords : Modeling, phenotypic traits, resistance

P-04.5
The role of wild ruminants as a reservoir of paratuberculosis in goat: a pilot study

Dezzutto D.[1], Barbero R.[1], Bergagna S.[1], Di Blasio A.[1], Maurella C.[1], Dondo A.[1], Quasso A.[2], Gennero M.S.[1]

[1]*Veterinary Medical Research Institute for Piemonte, Liguria and Valle D'Aosta (IZSTO) - Torino, Italy .* [2]*Veterinary Public Health (A.S.L. Asti) Asti, Italy*

Paratuberculosis (PTB) is a systemic infection and chronic inflammation of the intestine caused by Mycobacterium avium subsp. Paratuberculosis. PTB is now considered a cosmopolitan disease that affects a wide spectrum of species, including domestic ruminants and not. In small domestic ruminants, especially in goats, paratuberculosis have a strong impact on herd. Clinical symptoms are characterized by a massive decrease in milk production and a rapid progressive emaciation and cachexia (Windsor 2015).The purpose of this project is to analyze the trend of PTB in goats herds used for the production of Robiola di Roccaverano DOP cheese.The prevalence of paratuberculosis will evaluate in goats herds members of the Consortium of the Protection of the cheese. A serological survey and the isolation of mycobacterium from fecal samples will be performed on adults goats. In addition, it will be investigated the epidemiological situation of health status of roe deer (Capreolus capreolus) that lives in the same areas. During some planned roe deer culls we will collected blood samples and other biological matrices for the evaluation of paratuberculosis in wild ruminants. We will perform serological tests and the isolation from a representative sample of 200 animals. In addition, roe deer found dead during the whole period of the project will be necropsied.The results will be used to develop risk analysis aimed to assess the exposure to PTB for goats during the pasture period. Therefore, the results obtained will provide a valuable aid to farmers as regards the health and the livestock biosecurity management. Finally, since they were conduced very few studies on paratuberculosis in goats, the implementation of scientific data will improve the management of PTB in this species.bibliographyWinsor P.A. (2015) Paratuberculosis in sheep and goats, Veterinary Mocrobiology, Vol. 181, 1-2,pp. 161–169.

Keywords :
Paratuberculosis, goats, wild ruminants

P-04.6
Calf-to-calf transmission of MAP infection between calves born to MAP-positive and MAP-negative dams

Agnieszka Wiszniewska-Laszczych[1], Joanna Szteyn M.S.[1]

[1]*Department of Veterinary Protection of Public Health, Faculty of Veterinary Medicine, University of Warmia and Mazury in Olsztyn, Poland*

Objective : Define possibility of calf-to-calf transmission of MAP infection between calves born to MAP-positive and MAP-negative dams placed in the same stable.MaterialFaeces of 66 (15 from MAP-positive dams and 51 from MAP-negative dams) calves collected 3 times at 2-5, 180-185 and 360-365 days of age. After 30 days of life all animals were kept together in young stock group.
Methods : All faecal samples were evaluated for the presence of MAP by the cultivation method and direct isolation of DNA-MAP from the samples with Genomic mini kit (A&A Biotechnology, Poland). Colonies typical for MAP and DNA-samples were confirmed by PCR results and reported as positive or negative.
Results : Faecal samples of calves in the age of 2-5 days have not demonstrated positive results both by the cultivation method and direct isolation of DNA-MAP. In the age of 180-185 days 66.66% of calves from MAP positive and 13.72% calves from MAP-negative dams demonstrated at least one MAP-positive result. In the age 360-365 days MAP-positive results demonstrated respectively 73.33% and 15.68%. All 180-185 day-old MAP-positive calves were also MAP-positive in the age of 360-365 days. In 3 cases, 180-185 day-old calves with positive results only in direct isolation of DNA-MAP from faecal samples were also MAP-positive in the cultivation method in the age of 360-365 days. In 3 cases, 180-185 day-old calves with positive results only in the cultivation method were also MAP-positive in direct isolation of DNA-MAP from faecal samples in the age of 360-365 days. Calves with at least one positive faecal sample result were classified as MAP-positive.
Conclusion : In a herd with low MAP infection, both calves aged 6 and 12 months coming from MAP-negative and MAP-positive dams shed MAP in faeces.Calves coming from MAP-positive dams shed MAP with faeces much more often compared to calves originating from MAP-negative dams. The difference is statistically significant.In the research conducted during the 12th month of the life of calves, the number of animals shedding MAP with faeces increased, confirming the possibility of calf-to-calf transmission.

Keywords :
Calf-to-calf transmission, cultivation, DNA, direct isolation

P-04.7

BIO-PRESENCE AND MOLECULAR EPIDEMIOLOGY OF MYCOBACTERIUM AVIUM SUBSPECIES PARATUBERCULOSIS IN CAPTIVE POPULATION OF ZOO ANIMALS OF PUNJAB (INDIA)

Abdul Qayoom Mir[1], Kirti Dua[1], **Shoor Vir Singh**[2], Shukriti Sharma[1], Gursimran Filia[3], Mahendra Pal Singh[4]

[1]Department of Veterinary Medicine, COVS, GADVASU, Ludhiana-141 004 Punjab, India. [2]Microbiology Laboratory, Animal Health Division, Central Institute for Research on Goats, Makhdoom, PO- Farah – 281 122, Mathura, Uttar Pradesh, India. [3]Animal Disease Research Centre, COVS, GADVASU, Ludhiana-141 004 Punjab, India. [4]Mahendra Chaudhary Zoological Park, Chhatbir, Punjab, India

Mycobacterium avium subspecies paratuberculosis (MAP), the etiological agent of Johne's disease has emerged as a major animal pathogen of global importance with substantial zoonotic, socio-economic and public health concerns. MAP is endemic in the domestic livestock population of the country and has frequently been reported from livestock in Punjab. This study first time investigated bio-presence and genetic diversity of the MAP in the captive zoo animals of Punjab. A total of 159 fecal samples collected from 31 species of animals (15 herbivores species, 6 carnivores species, 7 primate species and 1 species of rodent from 5 different zoological parks of Punjab) were screened by three diagnostic tests for the detection of MAP, including microscopy, culture, IS900PCR and bio-typing using IS1311 PCR-REA. In microscopy 88 (55.3%) fecal samples were positive for AFB in distinguishable to MAP with [1+, 53 (60.2%); 2+; 25 (28.4%); 3+, 4 (4.5%) and 4+, 2 (2.2%)]. PCR analysis of the samples using IS900 showed three samples as positive (two from Hyenas and one from Jungle cat) for MAP. On further bio-typing of MAP DNA using IS1311 PCR_REA all the three DNA proved to be of 'Indian Bison Type', the dominant bio-type infecting livestock population of country. These positive samples had presence of acid fast bacilli in fecal microscopy (2+) although they were negative on culture. Of the total 159 fecal samples cultured on HEY medium with mycobactin J, 74 (46.5%) samples exhibited presence of cell wall deficient (CWD) colonies after 1 year of culture but no culture with typical MAP morphology was observed. In earlier studies on many occasions, on further incubation these CWD colonies have been found to develop cell wall. Animals positive in PCR were re-sampled after 8 months and were clinically normal were negative in fecal PCR although they were positive on microscopy (2+). The present study suggests the possibility that captive zoo animals act as reservoirs with intermittent shedding of MAP.

Keywords : Mycobacterium avium paratuberculosis, captive wild animals, Indian Bison type herbivores, carnivores

P-04.8

RISK FACTORS ASSOCIATED WITH MYCOBACTERIUM AVIUM SSP. PARATUBERCULOSIS HERD STATUS DETERMINED BY ENVIRONMENTAL CULTURE IN QUÉBEC DAIRY HERDS

Maria Puerto Parada[1], Elizabeth Doré[1], Julie Paré[2], Geneviève Côté[3], Juan Carlos Arango-Sabogal[1], Vincent Wellemans[1], Sébastien Buczinski[1], Jean-Philippe Roy[1], Olivia Labreque[4], Gilles Fecteau[1]

[1]Département de sciences cliniques, Faculté de médecine vétérinaire, Université de Montréal, Saint-Hyacinthe, Québec, Canada, J2S 8H5. [2]Agence canadienne d'inspection des aliments, Saint-Hyacinthe, Québec, Canada, J2S 7C6. [3]Direction générale des laboratoires d'expertise, Ministère de l'Agriculture, des Pêcheries et de l'Alimentation du Québec, Québec, Canada, G1P 4S8. [4]Laboratoire d'épidémiosurveillance animale du Québec, Ministère de l'Agriculture, des Pêcheries et de l'Alimentation du Québec, Saint-Hyacinthe, Québec, Canada, J2S 7X9

The importance of the control of paratuberculosis is justified by the associated economic losses and the potential role of Mycobacterium avium ssp. paratuberculosis (MAP) in human Crohn's disease. Control programs based on good rearing calves' practices are more effective than a simple test-and-cull strategy. The objective of this case-control study was to identify management practices significantly associated with MAP herd status determined by environmental culture. Herds were selected from the participants in the Québec Paratuberculosis Voluntary Control Program. A risk assessment questionnaire was administered to the owners. A total of 31 cases (herds from which MAP has been isolated from the environment) and 91 controls (herds that never reported having MAP infection and for which two consecutive yearly environmental samplings were negative) were included. Culture of MAP was achieved using the liquid media in the BACTEC 960 detection system. Potential risk factors associated with MAP herd status were analyzed in a univariate analysis and included in the multivariate analysis if $P<0.2$ and if they were considered biologically important in MAP transmission.Risk factors significantly associated with MAP status were: herd size, number of purchased cows, frequency of housing more than 1 cow in the maternity pen, hygiene conditions of the calving area, use of the calving area for another purposes, and allowing calves to nurse their dam.Logistic regression was used to evaluate the association between selected risk factors and MAP herd status. Final model was built following a forward stepwise approach. Frequency of housing more than one cow in the maternity pen was significantly associated with a positive MAP status (OR=3.33; 95% CI:1.06-10.49). Also, herds purchasing a higher number of cows in the last 5 years tended to be MAP positive (OR=1.03; 95% CI:1.00-1.06).The risk factors identified in the present study were related to the contact of newborn calves with adult animals or their feces and the introduction of new animals into the herd. Management practices to improve those 2 components together are key elements to minimize MAP transmission and introduction into a herd, and should be prioritized in control programs.

Keywords : Risk factors, Johne's disease, herd status, environmental culture

NOTES

NOTES

NOTES

NOTES

SESSION 5

One health & Control plans

Mycobacterium avium ssp. paratuberculosis in the context of One Health; working towards optimal health for people, animals and the environment.

Lisa Waddell, Public Health Risk Sciences, National Microbiology Laboratory, Public Health Agency of Canada

Introduction

The concept of One Health has been around since the late 19th century when Dr. Virchow demonstrated that diseases could be passed between humans and animals and evolved in the mid-20th century with the advent of veterinary public health which is underpinned by the understanding that good animal health is important for good public health [1]. For the past decade there has been increasing acceptance from governments, international organisations, researchers, and public health that greater interdisciplinary collaboration is required for optimal health of humans, animals and the environment. There have been many success stories in dealing with emerging diseases that could have significant impact, for example the global effort to address the on-going risk of an emerging pandemic influenza [2]. There are many other zoonotic diseases that have a much lower public profile, but could also benefit from a One Health approach [2]. Theoretically any control program for a zoonotic disease should take a holistic approach to strategizing and consider the benefits to humans, animals and society, which is likely to lead to the optimization of cost-effective strategies with wider support and participation from different sectors.

The elasticity of the One Health framework could be a beneficial approach to Mycobacterium avium ssp. paratuberculosis (MAP) control as it provides a widely accepted and recognised vehicle to promote dialogue and collaboration across disciplines to solve zoonotic disease issues at a local to global level. The key here is recognising that different disciplines likely have different roles, but may also have similar considerations, goals and challenges. It may also mean that the One Health approach is strategizing together, but resultant activities still remain within disciplines. Given MAP is not an emerging public health threat with high public pressure, the first challenge will be starting the conversation about drivers, facilitators, barriers and future scenario planning.

In order to understand some of the drivers and barriers of employing a One Health framework to MAP, an evidence-informed risk profile was constructed from a series of studies aimed at gathering primary research and expert opinion on the zoonotic potential and public health importance of MAP. The risk profile follows the principles of risk analysis for food standards adopted by the United Nations [3] and is used to identify consistencies within the evidence, areas of knowledge saturation and where critical knowledge gaps exist. The risk profile aims to put the evidence into context for a risk manager to evaluate the acceptable level of risk and mitigation required. The following sections briefly describe the evidence and knowledge gaps for the five sections of a risk profile; hazard identification, hazard characterization, exposure assessment, risk characterization and risk mitigation.

Hazard Identification

The hazard, MAP, is a well-studied organism that has been shown to cause Johne's disease in ruminants for which there is currently no treatment or cure [4, 5]. This disease has significant animal health and economic implications for the industry. The zoonotic potential of MAP has been investigated for more than 30 years as a potential cause of Crohn's disease in humans and more recently as a potential cause of other human diseases such as diabetes mellitus type 1 and multiple sclerosis among others that are captured in a recent systematic review [6]. The evidence summarized in this systematic review highlights an epidemiological association between MAP and human disease, which was corroborated by the responses of topic specialists to a survey on this topic where 93% indicated MAP was likely a risk to human health [7].

Hazard Characterization

Most of the critical knowledge gaps prevent us from characterising MAP as there is little information published on its pathogenesis in humans and none on a dose-response relationship between level of exposure and development of disease in humans [6]. The small number of surveys examining hypothetically high exposure occupations such as farmers with Johne's disease on the farm or large animal veterinarians failed to find an association with Crohn's disease [6]. Thus, it is possible that future research on genetic or other determinants of susceptibility will help to define who is susceptible to MAP and enable progressive research on the zoonotic potential of MAP. However, at this time the hazard characterisation is incomplete.

Exposure Assessment

Even though it is not certain that MAP causes human disease, conducting an exposure assessment can be important for understanding the sources or routes of human exposure to MAP. The findings of a scoping review on this topic, complimented by a topic specialist survey highlighted a vast number of potential sources for humans and animals, figure 1 [8]. The research to date provides evidence of human exposure mostly at low levels through a number of food sources including pasteurised or unpasteurized dairy products and meat, consumption of water, and other environmental media and direct contact with infected ruminants or animals. Some hypothesized sources of MAP such as seafood and produce have not been evaluated in the literature. Thus, there are important knowledge gaps in terms of the level of contamination and fluctuations in contamination of various human sources of MAP that prevent development of a plausible exposure assessment model.

Risk Characterization

Characterization of the risk MAP poses to humans can only be completed when we address knowledge gaps related to whether MAP is pathogenic to humans, who is the susceptible population and what exposure conditions are necessary to cause disease. Both the literature and the topic specialists surveyed agree that human exposure to MAP is likely at low rates from a variety of sources throughout an individual's lifetime [7, 8]. Fundamental challenges with studying MAP include the slow development of disease in animals and perhaps humans, low-levels of lifetime exposure from several common sources and the difficulty of identifying MAP in humans with Crohn's disease, animal samples or other media where the concentration of MAP is near the detection limits. These factors make understanding the epidemiology of this potential human pathogen extremely challenging.

Risk Mitigation

At this time risk mitigation efforts described by topic specialists are targeted at the farm level in an effort to control infection in ruminants [9]. The goal of the largely voluntary Johne's disease control programs is to prevent Johne's positive animals from entering a herd and minimise the risk of susceptible animals acquiring infection [10]. These biosecurity measures when combined with test and cull offer the most cost-effective mitigation options for most ruminants from an agri-food perspective [10, 11], and may also have the added benefit of reducing other pathogens on the farm. From a public health perspective, on-farm Johne's disease control programs should contribute to a decrease in the load of MAP in many sources of human exposure identified in figure 1. However, the impact of these mitigation programs on contamination of the environment and ruminant products has not been evaluated.

Most of the existing Johne's disease control programs for ruminants are voluntary, apply a risk-based testing scheme for milk samples (bulk tank or individual teat milk), ruminant samples (feces or blood) or pooled environmental fecal samples and have actions based on positive test results and the herd's historical MAP status [9]. Many countries cite spotty adoption of their programs by producers and difficulties in keeping their low or MAP-free herds enrolled in the program [9]. Some countries are offering subsidies for producer losses due to MAP, however in countries where subsidies existed and then were stopped, program compliance fell significantly. Global topic specialists considered that the current mitigation efforts in their country are sufficient (68.4%), while half thought additional resources are warranted, and the majority (92%) considered government's role is to keep apprised of this issue and not take a more active role [7]. At present there are some barriers to further optimization of the MAP control strategies including poor diagnostic test performance for early infection, costs associated with test and cull strategies and the lack of an ideal vaccine for all species [7].

Beyond farm level control programs, none were identified that targeted water or environmental contamination or any food for human consumption; however it is conceivable that interventions to minimize MAP could offer future marketing advantages [7, 9]. Similarly there were no existing consumer level mitigation or education programs identified by the topic specialists.

With respect to the zoonotic potential of MAP, there are some critical knowledge gaps in assessing the human health risk and public health impact. Thus, the current Johne's disease control programs were considered by the global topic specialists to be a proactive approach to controlling MAP, potentially decreasing human exposure while directly benefiting the ruminant industry through improved animal health and productivity [7]. Costs associated with these programs have been cited as a major barrier to wide-spread adoption as they are mainly paid by the producer and may include substantial initial investment in changes to the farm and costs of testing and premature culling of animals [7]. Economic assessments have projected these costs to be offset by decreasing cost of disease on the farm, better overall production, and may be further offset by marketing advantages for having a low-risk of Johne's disease herd [10, 11].

Conclusions:

Two points resonate from evaluating the literature and topic specialist's responses on this topic; first, we do not know the extent and type of role that MAP plays in human disease. Second, ruminants are the main known reservoir and no country or jurisdiction has increased its risk mitigation strategy beyond on-farm control programs. Most countries have encouraged development of voluntary on-farm Johne's disease control programs until additional evidence or superior control strategies are available [9, 10]. Producers and affected agricultural industries should be aware it is possible that future trade of ruminant derived products (semen, animals, dairy and meat) could require proof of MAP-free status [12]. Industry should be encouraged to continue to build Johne's disease (MAP) control programs that will have wide-spread benefit for the immediate herd as well as reducing MAP contamination of food and the environment. Overall, the current efforts should be seen as preventative measures that offer a degree of precaution with respect to public health and will also reduce the likelihood of future trade barriers due to MAP. It is prudent to continue monitoring and evaluating new evidence on MAP and periodically reassess to what extent the critical knowledge gaps have been addressed and whether there is sufficient evidence to assess the impact of MAP on public health. Given the ever increasing complexity of our food system and the many stakeholders that both contribute to and rely on the provision of safe and sustainable food, it is imperative that we strive to collaborate across disciplines (research, medicine, agriculture, environment and public health) and jointly strategize how to have an impact while efficiently using scarce resources.

References:

1. IOM (Institute of Medicine). *Improving food safety through a One Health approach.* Washington, DC: The National Academies Press., 2012.
2. Gibbs EP. The evolution of One Health: a decade of progress and challenges for the future. The Veterinary record 2014; 174: 85-91.
3. World Health Organization. Application of Risk Analysis to Food Standards Issues. The report of the Joint FAO/WHO Expert Consultation, 13–17 March 1995 1995; WHO/ FNU/FOS/95.3.
4. Gould, et al. ILSI Europe Emerging Pathogens Task Force. Mycobacterium avium ssp. paratuberculosis (MAP) and the food chain. FPT special report. Food Protection Trends 2005; April: 268-296.
5. Nacy C, Buckley M. Mycobacterium avium paratuberculosis: infrequent human pathogen or public health threat? Washington, DC: American Academy of Microbiology, 2008.
6. Waddell L, et al. The Zoonotic Potential of Mycobacterium avium ssp. paratuberculosis: Systematic review and meta-analyses of the evidence. Epidemiology and Infection 2015 Nov; 143: 3135-3157.
7. Waddell L, et al. The potential public health impact of Mycobacterium avium ssp. paratuberculosis: Global opinion survey of topic specialists. Zoonoses and Public Health 2016 May; 63: 212-222.
8. Waddell L, et al. Evaluating Pathways of Human Exposure to Mycobacterium avium ssp. paratuberculosis: A Scoping review of the existing knowledge [dissertation]: University of Guelph, 2016, 41 pp.
9. European Food Safety Authority. Proceedings of the 4th ParaTB Forum. In: Anonymous , Parma, Italy edn, 2014.
10. Garcia AB, Shalloo L. The economic impact and control of paratuberculosis in cattle. Journal of dairy science 2015; 98: 5019-5039.
11. Bennett R, McClement I, McFarlane I. An economic decision support tool for simulating paratuberculosis control strategies in a UK suckler beef herd. Preventive veterinary medicine 2010; 93: 286-293.
12. Momotani E. Epidemiological situation and control strategies for paratuberculosis in Japan. The Japanese journal of veterinary research 2012; 60 Suppl: S19-29.

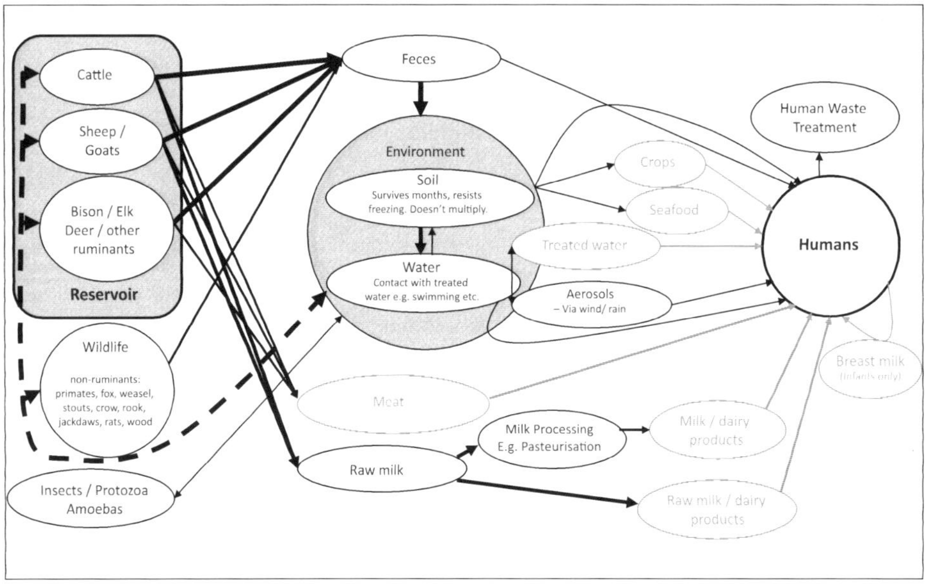

Grey: represents exposure pathways related to human consumption of water and food. Increasing thickness (———) of the lines that link sources indicates increasing amount of evidence and confidence that MAP contamination occurs along this route.

Figure 1: Potential pathways of human exposure to MAP weighted by the quantity and consistency of the current literature and topic specialists' opinion.

ORAL

O-05.1
MODELING PARATUBERCULOSIS IN A TYPICAL SHEEP FLOCK IN NEW ZEALAND

Marquetoux N.[1], Mitchell R.[2], Wilson P.[1], Stevenson M.[3], Ridler A.[1], Heuer C.a[1]

[1] *EpiCentre, Institute of Veterinary, Animal and Biomedical Sciences,Massey University,New Zealand.*
[2] *Department of Population Medicine and Diagnostic Sciences, College of Veterinary Medicine, Cornell University,USA.*
[3]*Faculty of Veterinary and Agricultural Sciences, The University of Melbourne, Parkville Victoria 3052 Australia*

Pastoral sheep farming in New Zealand (NZ) is characterized by all-year grazing and strong seasonality of production cycles. Infection with Mycobacterium avium subspecies paratuberculosis (MAP) is endemic in NZ sheep flocks with a farm-level infection prevalence of 79%. MAP is the causative agent of ovine Johne's disease (OJD) causing production loss in affected flocks.
To estimate the production loss and the financial consequences of OJD control, we developed a mathematical state-transition model for MAP transmission dynamics in a typical sheep flock. We first estimated transmission parameters by conducting a meta-analysis of natural and experimental infection studies of sheep. The resulting parameters informed a continuous time state transition model of OJD infection dynamics. Seasonal events of lambing, culling and replacement, their impact on annual offtake rates and associated financial outputs were incorporated in the demographics part of the model. Sheep newly infected with MAP could enter either a track leading to clinical disease or a track leading to recovery, as suggested by the meta-analysis. We thus evaluated the economic loss due to paratuberculosis and the cost-effectiveness of vaccination over time. One criterion was the minimum clinical OJD incidence above which vaccination was rendered profitable and the time required for reaching this break-even point. We also evaluated the impact of purchasing MAP infected replacement lambs on the effectiveness of vaccination.
The model represents a robust framework for infection dynamics of OJD. The economic model component may be used by farmers as a decision support tool for the management of OJD. Preliminary results suggested a threshold of 1% annual clinical OJD incidence above which profits from vaccination reached the financial break-even point after about 4-6 years of vaccinating lambs at 2-3 months of age.

Keywords :
Modeling ovine paratuberculosis, production effects, within-flock infection dynamics, cost-effectiveness of vaccination

O-05.2
SPREAD AND CONTROL OF BOVINE PARATUBERCULOSIS IN AN ENZOOTIC CATTLE REGION: A MULTI-SCALE MODEL TO EVALUATE COMPLEX STRATEGIES COMBINING BIOSECURITY AND TEST-AT-PURCHASE

Gaël Beaunée[1,2], Elisabeta Vergu[2], Alain Joly[3], **Pauline Ezanno**[1]

[1]*INRA, ONIRIS, UMR1300 BioEpAR, CS40706, F-44307 Nantes, France.* [2]*INRA, UR1404 MaIAGE, F-78350 Jouy-en-Josas, France.*
[3]*Groupement de Défense Sanitaire de Bretagne, 56019 Vannes, France*

Bovine paratuberculosis is mainly spread between herds due to trade movements of infected and undetected animals The prevalence worldwide being high at animal and herd levels, and infected animals being hard to detect using routine diagnostic tests, the disease spread cannot be easily observed in the field, whereas there is a need for assessing control strategies. Our objective is to better understand the spread of Mycobacterium avium subsp. paratuberculosis (Map) at a regional scale using a modelling approach, and to compare through intensive simulations complex control strategies combining biosecurity measures (early culling, hygiene improvement, calf management) and tests-at-purchase.
We developed the first multi-scale mechanistic model of Map spread between dairy cattle herds, accounting for stochastic within-herd dynamics (demography and infection), indirect local transmission, and incorporating data on animal trade and on herd-specific size and management. We modeled all of the 12,857 dairy herds located in Brittany (France) having more than 15 dairy females. Data from 2005 to 2013 was used to calibrate each herd size and demographic rates, and to define trade events. We assumed initially 30% of the herds to be infected with a 10% within-herd prevalence on average.
Each measure tested alone or in combination with tests at purchase succeeded in slowing down the regional Map spread, but not in decreasing the proportion of infected herds. More than two measures had to be combined to effectively reduce the herd-level prevalence. In such a case, only a moderate level of implementation of each measure was required, indicating the operational potential of such combined strategies.
Our study highlights the challenge of controlling Map spread in an endemically infected region because of poor test characteristics and frequent trade movements. Our model is a flexible and efficient tool to help collective animal health managers in defining relevant control strategies at a regional scale, accounting for regional specificities in terms of contact network and farm characteristics.

Keywords :
Multi-scale modelling, complex control strategies, dairy cattle, intensive simulations, biosecurity, trade

O-05.3

CHANGES IN PREVALENCE OF OVINE PARATUBERCULOSIS FOLLOWING EXTENDED VACCINATION WITH GUDAIR OVER MORE THAN A DECADE

Navneet Dhand[1], Peter A. Windsor[1], Richard J Whittington[1]

[1]*The University of Sydney, Australia*

Gudair™ vaccine was registered in Australia in 2002 for control of ovine Johne's disease (OJD) and has since become the key tool for management of this disease. As an understanding of the long term effectiveness of vaccination is important for strategic national disease control, we report on a study that determined changes in prevalence of shedding over more than a decade in a cohort of 12 OJD infected flocks that commenced vaccinating lambs in or prior to 2002.
Flocks with variable initial prevalence (5, 4 and 3 with low, medium, and high prevalence, respectively) were enrolled in the study in 2003-04. Six biennial fecal samplings were conducted in these flocks. At each sampling, we aimed to select 7 or 14 faecal pools of 25 or 50 sheep each (collecting one pellet per sheep) from each of the four age groups (3, 4, 5 and 6 year-old-sheep) from each flock, although the actual numbers and sizes of pools did vary. Samples were cultured using pooled faecal culture and the sheep level OJD prevalence was calculated. Changes in probability of a pool to be positive and in OJD prevalence were evaluated by fitting linear- and generalised-linear mixed models.
The proportion of positive pools significantly declined over time from 50.3% at the first sampling in 2003-04 to only 3.1% at the last sampling, suggestive of a 30 fold reduction in the odds of a pool to be positive ($p<0.001$). Similarly, the average animal level prevalence dropped from 7.64% at the first sampling to just 0.12% at the last sampling. However, 7 of the 10 flocks remaining in the project at the second last sampling conducted in 2011-12, and 3 of the 8 flocks at the last sampling in 2013-14 had sheep with detectable shedding of MAP after more than a decade of vaccination. The results confirm that Gudair™ vaccination is effective in reducing OJD prevalence but MAP shedding persists at low-level in vaccinates in some flocks even after extended vaccination. Farmers should carefully evaluate the risk of introduction of MAP to their farms when purchasing vaccinated re-stocker sheep, particularly from known or suspected OJD infected properties.

Keywords :
Ovine Johne's disease; Gudair™; Vaccination; Abattoir surveillance; Pooled faecal culture; Paratuberculosis control programs

O-05.4

CHANGES IN PREVALENCE OF OVINE PARATUBERCULOSIS IN SPECIFIC COHORTS OF VACCINATES OVER FIVE, TWO-YEAR INTERVALS

Jeffrey Eppleston[1], Navneet Dhand[1], Peter A. Windsor[1], Richard J Whittington[1]

[1]*The University of Sydney, Australia*

A longitudinal field trial was conducted in twelve flocks to determine changes in prevalence of ovine Johne's disease (OJD) after commencing long term vaccination with Gudair™. Flock-level prevalence changes due to vaccination are reported elsewhere at this colloquium. In this paper we present changes in prevalence in age-specific cohorts measured at two-year intervals for more than 10 years.
Faecal samples were collected on six occasions at 2 year intervals when we aimed to sample up to 350 sheep from each of four age-specific cohorts (3, 4, 5 and 6 year-old). Samples were cultured using pooled faecal culture in pools of 25 or 50 and the sheep level OJD prevalence calculated. At the first sampling only 1 and 2 year-old sheep had been vaccinated so that all sheep at the first sampling and 5- and 6-year-olds at the second sampling were not vaccinated. Since sampling was conducted every two years, the 3 and 4 year-old sheep were next sampled when they were 5 and 6 year-old, respectively.
Descriptive and linear mixed model (LMM) analyses were conducted to evaluate changes in prevalence in cohorts followed over two consecutive samplings. In LLM analyses the effect of time (first or second sampling within cohort), cohort (1 to 5 representing sampling years), and age group ('3- to 5-year-old' or '4- to 6-year-old sheep') on log prevalence was evaluated, using flock as a random effect.
Descriptive analysis showed no consistent pattern in animal level OJD prevalence when cohorts of sheep within flocks were followed for two successive samplings with some falling, some steady and some increasing suggesting that the effect of vaccination on prevalence is highly variable between flocks. Reasons for this variation could be related to variations in management, including introduction of infected sheep, presence of super-shedder sheep and variable biosecurity. Year of sampling (cohort) was the only significant variable in LMMs suggesting that the prevalence dropped significantly over five samplings. Prevalence was not significantly different between two consecutive samplings in the same cohort of sheep within flocks suggesting that vaccination over long term is required to make an impact.

Keywords :
Ovine Johne's disease, Gudair™, Vaccination, Pooled faecal culture, Paratuberculosis control programs

O-05.5
ECONOMIC EFFECTS OF JOHNES DISEASE AND ESTIMATES OF VACCINE EFFICACY ON SHEEP FARMS IN NEW ZEALAND

Milan Gautam[1], Peter Anderson[2], Anne Ridler[1], Cord Heuer[3]

[1]Institute of Veterinary, Animal and Biomedical Sciences, Massey University. [2]The Vet Centre Marlborough, Blenheim. [3]EpiCentre, Institute of Veterinary, Animal and Biomedical Sciences, Massey University, New Zealand

Seventeen conveniently selected sheep farms were enrolled from the South and North Islands of New Zealand and followed up for up to two years. The primary objective was to investigate production loss attributable to ovine Johnes disease (OJD) mortality. An additional aim was to estimate the likely financial benefit from vaccinating lambs against OJD and the circumstances required for positive returns. The total study size was 29 farm-years.
Farms were categorized as fine wool (Merino, Half-bred, Corriedale) and other breeds (Romney, composite breeds). OJD was confirmed by gross- and histo-pathology and ELISA examination of 390 ewes selected for culling due to chronic progressive wasting. A sample of live ewes with poor body condition score (BCS) was also tested to estimate the extent of OJD in ewes in the lower 10% BCS of the flock.
The overall ewe mortality was similar in fine wool vs. other breeds (7.3 vs. 8.0%; farm range 2.8-15.7%), the estimated annual OJD mortality was four-fold as high in fine wool (2.8%, farm-adjusted inter-quartile range IQR 1.9-4.0) as in other breeds (0.7%, IQR 0.4-1.2) with large variation between farms. ELISA results of live ewes also demonstrated fine wool sheep had a higher predisposition to OJD (34.8%, STD=30.8%, n= 101 vs. 6.85%, STD=8.1%, n= 96). Stochastic modelling indicated that the cost of mortality per ewe was $9.4 (IQR 6.5 – 13.0) in fine wool and $2.3 (IQR 1.4 – 3.6) in other breeds. Vaccinating fine wool breeds against OJD may be cost-effective in most flocks when the annual clinical OJD loss is over 1/100 ewes. Since annual OJD incidence was lower in ewes of other breeds, only marginal financial gains may be derived from vaccinating other breeds against OJD. Vaccination may be advisable for farms where the OJD incidence is consistently higher than the 1% threshold. To make a rational decision about vaccination, it is therefore critical to monitor and record OJD among ewes.

Keywords :
OJD, all year grazing, mortality, production cost, vaccination

O-05.6
NEW CRITERIA TO DIFFERENTIATE PARATUBERCULOSIS VACCINATION FROM BOVINE TUBERCULOSIS INFECTION IN THE CERVICAL COMPARATIVE SKIN TEST

Serrano M.[1], Elguezabal N.[1],Urkitza A.[2], Geijo M.V.[1], Molina E.[1], Arrazuria R.[1], A. Sevilla I.[1], Vordermeier M.[3], Whelan P.[3], Juste R.A.[1,4], Garrido J.M[1]

[1]NEIKER - Tecnalia, Basque Institute for Agricultural Research and Development, Derio, Bizkaia, Spain. [2]Veterinary Surgeon, Gernika, Bizkaia, Spain. [3]Department of Bovine Tuberculosis. Ahvla, Surrey, United Kingdom. [4]SERIDA, Department of Agriculture of the Regional Government of the Principality of Asturias, Deva, Asturias, Spain

Although the efficacy of paratuberculosis (PTB) vaccination has been repeatedly demonstrated, its use has been restricted due to its interference with current tuberculosis (TB) diagnostic methods used in official eradication schedules. The goal of this study was to assess new differentiation of PTB vaccination from bovine TB infection in the Cervical Comparative Skin Test (CCST).
Fifteen Friesian calves were selected from a feedlot based on IFN-γ release assay (IGRA) negative results through 3 consecutive samplings at weeks 0, 4 and 12 (W0, W4, W12). Ten of them were vaccinated with 1ml of an inactivated PTB vaccine at W0. After introduction into level 3 biosafety containment facilities, 5 vaccinated and 5 non-vaccinated calves were infected with 105 CFU of M. bovis by the intratracheal route on W16, resulting in 3 different groups: Map vaccinated-M. bovis infected, only Map vaccinated and only M. bovis infected. The CCST was carried out at W13 with Avian (APPD) and Bovine PPD (BPPD).
Validation of the new criteria for Sensitivity (Se) was carried out in a herd of 120 Friesian cows unknowingly infected with M. bovis when vaccinated against PTB with a commercial vaccine (Silirum) and annually submitted to CCST during 3 years. Validation for Specificity (Sp) was carried out in over 9000 CCST on PTB vaccinated cattle from 26 herds throughout 9 years. Results were read and interpreted according to the standard criteria based on the absolute differences of increase in mm between both PPDs and an alternative one based on the calculation of the relative increase in percentage of each one (>100% and BPPDfflAPPD).
In the experimental setting, the standard interpretation yielded a Sp of 100%, but a Se of 40% for vaccinated animals. Using the alternative criteria, Se reached 80% and Sp 100%. In field conditions, standard interpretation resulted in a Se of 23.9% and a Sp of 99.8% while the relative interpretation yielded a Se of 81.7% and a Sp of 99.8%.
These results indicate that the reduction of Se in the CCST caused by PTB vaccination can be advantageously overcome with the relative interpretation.

Keywords :
OJD, all year grazing, mortality, production cost, vaccination

O-05.7

THE THURINGIAN BOVINE JOHNE'S DISEASE CONTROL PROGRAM – RESULTS AND CONCLUSIONS FOR THE FUTURE CONTROL STRATEGY

Karsten Donat[1], Heike Köhler[2]

[1]*Thueringer Tiersenchenkasse, Animal Health Service, Jena, Germany.* [2]*Institute of Molecular Pathogenesis, Friedrich-Loeffler-Institut, Federal Research Institute for Animal Health, Jena, Germany*

Control of Johne's disease is being discussed controversially. Prevalence reduction had been demonstrated, but prospects to keep herds free from Mycobacterium avium ssp. paratuberculosis (MAP) and the sustainability of the control efforts are questioned. This study aimed at evaluating the results achieved by voluntary paratuberculosis control measures in cattle herds in Thuringia, a federal state of Germany, between 2008 and 2014, and at drawing conclusions for further practice.
A total of 76 dairy herds and 29 cow-calf beef herds, keeping a total of about 30,000 cows were included into the analysis. Cumulative incidence (CI) was used to monitor the control progress; new cases were detected by means of annual testing of the herd's cows by individual faecal culture. Herds with at least one positive test result were classified as MAP-positive. Herds without any detection of MAP during three consecutive years were certified as MAP non-suspect. Compared to the MAP-positive ones, herds tested negative at the beginning of the program had a higher chance (odds ratio 32.9; $p < 0.0001$) of achieving this certification by 2014. Thirteen out of 67 initially MAP-positive herds (19.4%) were certified according to the control programme. In a subset of 25 MAP-positive dairy herds that had been involved since 2008, CI decreased significantly from 14.0% to 5.6% in 2014. Regarding the initial situation in 2008, control progress was significantly higher in herds with CI > 5% compared to herds with CI < 5% as shown by two-way ANOVA.
The results support the hypothesis that eradication of paratuberculosis is feasible at herd level. A herd monitoring based on faecal culture tests and a certification period of three years seems to be adequate to justify the status MAP non-suspect, even if the herd had been MAP-positive in the past. From the results can be assumed that the status can be maintained by implementing biosecurity and trade controls. Once herds achieve a low level of CI, control efforts should be intensified.

Keywords :
Regional control program, herd certification, faecal culture

O-05.8

MILK QUALITY ASSURANCE PROGRAMME FOR PARATUBERCULOSIS: DECREASING PROPORTION OF ELISA-POSITIVE HEIFERS

Weber MF[1], Heuer C[2], Aalberts M[1], Schukken YH[1]

[1]*GD Animal Health, Deventer, the Netherlands.* [2]*EpiCentre, Massey University, Palmerston North, New Zealand*

In 2006, a milk quality assurance programme (MQAP) for paratuberculosis in Dutch dairy herds was initiated. The aim of the MQAP is to reduce the concentration of Mycobacterium avium subsp. paratuberculosis (Map) in milk delivered to the milk factories. Herds participating in the MQAP are assigned a herd status based on the results of herd examinations by individual milk-ELISA of all lactating cattle. Farmers are entitled to confirm positive ELISA results by faecal culture. Test-negative herds are assigned status 'A'. Test-positive herds are assigned status 'B' (if all test-positive cattle have been removed from the herd) or status 'C' (if any test-positive cattle are retained). Intervals between subsequent herd examinations are two years in herds with status 'A', and one year in herds with status 'B' or 'C'. The MQAP promotes preventive management measures and culling of test-positive cattle to reduce the spread of Map.
The aim of this study was to evaluate the effect of the programme on the spread of Map within the first cohort of 718 herds that voluntarily entered the MQAP in 2006 – 2007, without prior participation in a paratuberculosis programme. The apparent ELISA prevalence of heifers was used as a proxy-parameter for the spread of Map. By restricting the analysis to the first test result of individual animals, a direct effect of culling test-positive individuals on the course of the apparent prevalence was eliminated.
At the initial assessment (Year 0), 0.5% of the heifers had a positive milk ELISA result (i.e., S/Pffl1.00). In years 1 to 7, this proportion fluctuated between 0.3% and 1.0%. This strong fluctuation was related to the different test intervals between herd statuses as only herds with positive test results at the initial assessment were obliged to perform a herd examination in year 1, whereas both positive and initially negative herds were re-tested in year 2. After year 7, only ffi0.2% of heifers were ELISA-positive, indicating that the MQAP positively contributed to the control of Map in Dutch dairy herds.
In this paper, progress achieved in this cohort of 718 herds in a ten year period (2006 – 2015) will be presented.

Keywords :
Paratuberculosis, dairy cattle, programme

O-05.9

Implementation of control measures for bovine paratuberculosis in Swiss dairy and beef herds

Mireille Meylan[1], Myriam Anderegg[1]

[1]*Clinic for Ruminants, Vetsuisse Faculty, University of Berne, Switzerland*

Background: The most important risk factors associated with high prevalence of shedding of Mycobacterium avium subsp. paratuberculosis (MAP) have been recently described for Swiss dairy and beef operations (Künzler et al., 2014, BMC Vet Res 10:132).
Objectives: To monitor farmers' compliance in dairy and beef herds infected with paratuberculosis for the implementation of control measures to reduce MAP shedding.
Materials and Methods: After detailed assessment of each farm regarding management and risk factors for MAP transmission, the managers of 17 herds (11 dairy and 6 beef herds) received specific recommendations for measures to reduce MAP transmission and so better control the disease. The herds were visited twice a year during 3 years to follow the implementation of the recommended measures. After 3 years of observation, a detailed final assessment on the implementation of control measures was performed, and the compliance and perseverance of the herd managers with the implementation of these measures were evaluated.
Results: The most common recommendations regarded calving management, separation of adult animals from the calves (dairy) and heifers (dairy and beef), culling of animals shedding MAP and their offspring, and hygiene on the farm (including manure and pasture management). A mean of 8.3 specific control measures were recommended per farm (8.7 in dairy and 7.5 in beef farms), of which a mean of 3.2 were realized (3.3 in dairy and 3.0 in beef farms). Easy-to-implement measures were mostly realized, none of the farmers made deeper changes, e.g. modification in the buildings used for cattle. While cows shedding MAP were culled in all herds, their calves remained in the herd in 12 of the 17 farms. Implemented measures were generally kept in place for the entire study Duration.
Significance and conclusions: The recommended measures to control paratuberculosis were only partially (40%) implemented in infected herds. These results indicate that the losses associated with paratuberculosis in Swiss dairy and beef operations are not estimated by the farmers to be high enough to justify important efforts for control measures. This is in contrast to countries with larger herds and higher prevalence of the disease in infected herds.

Keywords :
Control, compliance, risk factors, dairy, beef

O-05.10

Shifting strategy for paratuberculosis management in Queensland

Lawrence Gavey[1]

[1]*Principal Veterinary Officer (Animal Disease Containment), Biosecurity Queensland, Department of Agriculture and Fisheries, PO Box 102, Toowoomba, Queensland, Australia*

Queensland has prevented and controlled spread of paratuberculosis into and within the State by regulation, including minimum entry requirements and movement restrictions when disease is confirmed or suspected. This approach has been supported by a presumed low prevalence of disease.
Detections of paratuberculosis in Brahman beef cattle in Queensland in 2012 and 2013 led to investigation and control responses involving 280 properties. Except for the index infected properties and four others retaining at-risk cattle in isolation, the risk statuses of all properties have been resolved by testing and/or culling. The key findings of these large incident responses are that paratuberculosis is difficult to detect in extensively grazed beef herds, the apparent prevalence of infection in the Queensland cattle industries is low, stud (seedstock) herds play a crucial role in managing risk of spread, and regulatory incident responses for paratuberculosis impose high financial and personal impacts on affected producers that exceed the perceptible nett benefits.
In response to these and other incidents, the Australian cattle industries are shifting national focus from regulatory protection of low-risk areas and herds to risk- and market-based industry management. New Queensland legislation identifies shared responsibilities for recognition and risk-based management of biosecurity threats, consistent with the new national framework. Consequent alleviation from the adverse impacts of regulatory control is expected to improve producer acceptance and compliance with obligations. National and State awareness programs are providing industry with the knowledge and tools needed for effective risk management.
This shift in focus is complemented by incorporation of a range of established diseases into integrated, whole-farm biosecurity planning to enhance productivity and supply-chain confidence. It follows successful piloting of industry management of paratuberculosis in the sheep industries.

Keywords :
Queensland, beef cattle, industry management, biosecurity

O-05.11
Success and failure in the control of Paratuberculosis in dairy herds by riskmanagement

Richard Sibley[1], Peter Orpin[2]

[1] *BVSc HonFRCVS, Westbridge Veterinary Group, Witheridge.* [2]*BVSc MRCVS Park Vet Group, Leicester, UK*

Dairy herds in the UK are offered six different strategic options for the prevention and control of Paratuberculosis. The most popular of these options is the management of high risk cows as identified by quarterly milk testing to prevent new infections. The rate of progress towards control and eradication is highly variable, and this paper will demonstrate the variation and explain why such a control strategy can be inappropriate in some herds leading to failure of control.
Method : 763 dairy herds have provided risk and test data to enable future prevalence to be predicted. 53% of these herds have risks and lack of control that indicate future prevalence is likely to be much higher. Only 6% of these herds had a declining predicted prevalence. Control in these high risk herds by the management of transmission is challenging, and in some cases, impractical and impossible. This study tracks progress in five herds that have predicted high risks of transmission, and looks at how these can, and cannot, be controlled in practice.
Results : The five herds show varied success of risk management as a strategic control of Paratuberculosis in the herds, as measured by positive milk test elisas and the age profile of test positive cows. Repeated reviews of transmission risks indicated that some major risks of transmission failed to be adequately managed in some herds, leading to high disease reproduction ratios and increasing prevalence. However, some herds managed to control risks of transmission from high risk cows that were identified by quarterly milk elisa testing very successfully, reducing prevalence effectively over five years.
Conclusions : The study indicates that quarterly milk testing and control of Paratuberculosis in dairy herds can be a very effective strategy for the control of Johnes disease in dairy herds. However, the strategy is inappropriate and ineffective in some herds that are unable to control some major specific risks, leading to an increased prevalence that is not evident until several years after controls are implemented.

Keywords :
Risk management, quarterly milk testing, control plan, predicted prevalence

O-05.12
Are we feeding live Mycobacterium avium subsp. paratuberculosis in calf milk replacer?

James C. Tarrant[1], Brenna Kunkel[1], Sheila McGuirk[1], Antonio C.G. Foddai[2], Irene R. Grant[2], Michael T. Collins[1]

[1]*School of Veterinary Medicine, University of Wisconsin-Madison, Wisconsin, USA.* [2]*Institute for Global Food Security, School of Biological Sciences, Queen's University Belfast, Northern Ireland*

When advising farmers on how to control Johne's disease, the #1 recommendation is to avoid feeding waste milk to calves and instead to feed them calf milk replacer (CMR). Obviously, this advice is based on the assumption that milk replacer is free of live Mycobacterium avium subsp. paratuberculosis (MAP) organisms capable of causing infection. No one has ever challenged this assumption. Preliminary work on CMR sourced in Wisconsin found that 1 (12.5%) of 8 samples tested positive for live MAP organisms by the peptide-mediated magnetic separation–phage assay (PMS-PA). Previously, 30 (44%) of 68 powdered milk products intended for human consumption were positive for live MAP by the same assay. The PMS-PA permits sensitive and rapid detection of live MAP. MAP cells are separated from the sample using magnetic beads coated with MAP-specific peptides and incubated with a mycobacteriophage, before treatment with viricide (to kill extracellular phages) and then plating with fast-growing M. smegmatis. Plaques in the resulting M. smegmatis lawn indicate the presence of live MAP organisms that have burst as a result of phage amplification. IS900 PCR is applied to plaques to confirm that the plaque is derived from MAP cell(s). We hypothesized that live MAP organisms may be widely prevalent in commercial powdered CMR products, so tested a broader sample of CMR products (n=83) obtained from dairy farms around the USA. In addition to the PMS-PA, conventional microbiological methods were used to quantify total mesophilic bacterial counts, coliforms, Salmonella, staphylococci, and streptococci in the CMR samples, in order to assess their overall hygienic quality. Sixteen (19.2%) of 83 CMR samples tested positive for live MAP based on PMS-PA with IS900 PCR confirmation of MAP DNA in plaques obtained. All conventional microbiology results were within the USA regulatory guidelines for bulk tank milk. No correlation was found between conventional microbiology and presence of MAP indicated by the PMS-PA. Corroboration of PMS-PA findings was attempted using sample processing with and without PMS, and solid and liquid medium-based MAP culture; culture results are pending. The implications of results of CMR testing for MAP control programs will be discussed.

Keywords :
Milk replacer,control,exposure

O-05.13
Johne's disease management in UK dairy herds participating in a quarterly testing programme - results of a farmer survey

Karen Bond[1]

[1]*National Milk Records, Royal Veterinary College, Carnforth, UK*

Participation in laboratory organised quarterly (milk antibody ELISA) Johne's disease testing schemes has increased dramatically over the last five years. Whilst this demonstrates greater awareness and engagement from farmers, we have little information about how these results are being used on farm, or if these farms have credible, robust management plans in place.
National Milk Records quarterly Johne's disease testing programme 'HerdWise' now has over 1500 herds participating. During 2015 & 2016 information has been gathered from 200 farmers enrolled on the HerdWise scheme through the completion of a questionnaire.
Results show large differences in the management of positive cows between farms and also in the treatment of 'amber' (single positive result) and 'red' (multiple positive results) cows within herds. Although all respondents have specific management plans in place for their 'red' cows 10% are doing nothing with their 'amber' cows. Management practices most commonly utilised for 'red' cows are immediate culling or segregation at calving and milk and colostrum discarding, but again showed wide variance between farm as to what was deemed important. Over 95% of respondents retained heifer calves from Johne's positive cows for breeding.
Veterinary engagement was also examined and again showed wide variance. Over half of respondents discussed each new set of quarterly results with their vet to decide on management of individual cows. However, 15% of farmers surveyed had never discussed Johne's management with their vet, despite the requirement of a veterinary signature at enrollment on the testing scheme.
These figures are based on analysis of the first 100 survey responses and will be updated for the final submission. These initial results demonstrate the need for veterinary engagement and continued education of farmers to ensure that testing is accompanied by a credible and robust management plan.

Keywords :
Johne's disease, paratuberculosis, management, risk, quarterly testing, survey, questionnaire

O-05.14
Motivating On-Farm Change for Johne's Disease Control Using Peer Learning and Whiteboard Videos

Steven Roche[1], Dr. David Kelton[1], Dr. Ann Godkin[2]

[1]*Department of Population Medicine, Ontario Veterinary College, University of Guelph.* [2]*Ontario Ministry of Agriculture, Food and Rural Affairs, Canada*

This study evaluated an experiential learning program, Focus Farms (FF), aimed at motivating adoption of on-farm management practices for Johne's disease (JD) control in the Canadian dairy industry. FF is a learner-centered process, which utilizes peer learning to facilitate behavioural change. Producers engage in 4 full-day sessions, where small groups collaboratively address their on-farm problems. Pre-post questionnaires collected data on 70 FF and 62 control respondents' knowledge, attitudes and behaviours; pre-post risk assessments were used to assess on-farm risk of JD transmission. Between 2010 and 2013, over 200 producers participated in FF. Eighty-one percent of FF respondents reported making at least one on-farm change; significantly more than that of control respondents (38%). Overall, FF respondents significantly lowered their risk score in 4/5 risk areas and had a significant reduction in their overall risk score between pre-post measurements. Qualitative investigation into perceived barriers inhibiting adoption uncovered physical (e.g. time, money, infrastructure) and intrinsic (e.g. priority, habits, motivation) barriers to change. Producers also perceived that many on-farm recommendations are impractical and/or ineffective. They further suggested that both extrinsic (e.g. incentives, penalties) and intrinsic (e.g. pride, responsibility) forms of motivation are necessary for widespread change. Informed with the results of the FF process, a novel online video series, 'Johne's Disease in Canadian Dairy Herds', was created for stakeholder education. These videos use a narrated script, cued to a series of time-lapse whiteboard drawings. This series addresses the perceptions discussed above, and communicates the importance of, and methods for, JD control. The first of these videos was presented successfully at the 12th ICP in Parma, Italy and is now available in 4 languages, with over 10,000 views on YouTube. The second video informs veterinarians about methods for motivating their clients to adopt changes for JD control. We aim to debut the third video at the 13th ICP in Nantes, France; this video will communicate to stakeholders about current JD research and international JD control efforts. Overall, the FF process is an effective method for motivating on-farm change, and the whiteboard series is a popular and effective means of engaging farmers online around the world.

Keywords :
Disease control, biosecurity, communication, peer learning, motivation, video

O-05.15
The use of a web based prevalence predictor and risk model to engage farmers in Johne's control

Peter Orpin[1,2], Richard J. Sibley[3]

[1]Park Vet Group, Whetstone, Leicestershire. [2]Special Lecturer, Nottingham University, Nottingham, United Kingdom, 3 Westridge Veterinary Group, Witheridge, UK

The success of Johne's disease (JD) control on a farm or regional level is largely defined by the degree of engagement that can be achieved with both the participating farmers and veterinarians. The occasional clinical case or test positive animal in a herd typically results in a reactive cull and may fail to trigger a rigorous examination of risks or creation of a robust JD control plan. In high-risk herds the true herd prevalence can rapidly accelerate unchecked. Controls are typically implemented when the disease has progressed and the prevalence is much higher complicating control at farm and regional level . In 2009 a web based prevalence prediction tool was developed (www.myhealthyherd.com) to help predict the future JD prevalence of infection within a herd. The prevalence predictor is part of an overall JD module within the myhealthyherd program that allows for comprehensive assessment of risks of entry and spread for the key infectious diseases of beef and dairy herds. These risks are managed by the creation of farm specific biosecurity, vaccination and control plans. A structured risk assessment of JD entry and spread risks is used which scores the risks using visual traffic light scoring and also listing and explaining the relevant risks. The risk profiles created are also used to create a predicted true herd prevalence for the herd based on the risk score and the test prevalence derived from either a targeted 30 cow milk ELISA screen or whole herd screen. The estimated true herd prevalence for the next 5 years is graphically displayed by multiplying the test herd prevalence with a factor dependent on the risks. Herds with high risk of spread can visualise the potential impact of these risks on their own future herd prevalence and are therefore prompted to take early action. The prevalence report created provides further decision support on the most suitable control strategy for the farm by selecting control strategies suited to the particular prevalence and risks. The use of web-based tools provide low cost options for education, effective engagement and management of Johne's disease.

Keywords :
Myhealthyherd, control, Johne's, paratuberculosis, risks

O-05.16
Dairy farmers' perceptions towards improving on-farm Johne's disease prevention and control

Caroline Ritter, Jolanda Jansen[1], Keliesha Roth[2], John P. Kastelic[2], Cindy L. Adams[3], Herman W. Barkema[2]

[1]Department of Production Animal Health, Faculty of Veterinary Medicine, University of Calgary, Calgary, AB, Canada, T2N 4N1. [2]St. Anna Advies, 6525 Nijmegen, The Netherlands. [3]Department of Veterinary Clinical and Diagnostic Sciences, Faculty of Veterinary Medicine, University of Calgary, Calgary, AB, Canada, T2N 4N1

Introduction: Implementation of management practices is currently regarded as the most effective way to reduce the prevalence of Johne's disease (JD) on dairy farms. However, farmers often fail to adopt recommended strategies to decrease transmission of Mycobacterium avium subspecies paratuberculosis (MAP), the cause of JD. The purpose of this study was to characterize farmers' perceptions towards JD prevention and control in general and recommended management strategies specifically.

Material and Methods: Qualitative research methodology (classical grounded theory) was used. Twenty-five farmers participating in a voluntary JD prevention and control program in Alberta, Canada were selected. A semi-structured interview approach was used and conversations were audio-recorded. Transcription and coding were done between interviews and emerging categories were described.

Results: Two main concepts distinguished farmers' perceptions. First, their belief in the importance of JD and second, their belief in recommended JD prevention and control strategies. Farmers were categorized into four groups according to these concepts: Proactivists, Unconcerned, Disillusionists, and Deniers. Proactivists and Disillusionists regarded JD as an important issue, whereas Proactivists and Unconcerned believed in proposed management strategies. Groups that regarded JD as more important were better informed about best practices to reduce MAP transmission. However, generally farmers did not regard JD control as a "hot topic" in communications with their veterinarian or other producers, and even farmers that received MAP-positive test results for their farm often did not consider JD an important problem.

Conclusion and relevance: Knowledge of the main concepts described in this study will enable stakeholders to understand farmers' perceptions and overcome barriers for implementation of JD prevention and control strategies. Improved communication should enhance development and success of voluntary JD programs.

Keywords :
Behavior, mindset, biosecurity, control program

POSTER

P-05.1

EVALUATION OF PRIME-BOOST IMMUNIZATIONS WITH A HEAT-KILLED MYCOBACTERIUM AVIUM SUBSP. PARATUBERCULOSIS VACCINE, AND MODIFIED VACCINIA ANKARA VIRUS EXPRESSING ANTIGEN 85A IN A MOUSE MODEL

Colombatti Olivieri[1], Maria Alejandra[1]; Moyano, Roberto Damian[1]; Montenegro, Valeria[1]; Mon, María Laura[1]; Viale, Mariana[1]; Alonso, Natalia[1]; Del Medico Zajac, Paula[1], Calamante, Gabriela[1]; Delgado, Fernando [2]; **Santangelo, María de la Paz**[1], Romano, María Isabel[1]

[1]*Instituto de Biotecnología, INTA Castelar, Buenos Aires, Argentina.* [2]*Instituto de Patobiología, INTA Castelar, Buenos Aires, Argentina*

INTRODUCTION: Paratuberculosis is a chronic disease caused by Mycobacterium avium subsp. paratuberculosis (Map). Considering the economic losses, it is necessary to control the disease and to develop an effective vaccine. To evaluate the protection conferred by a heat-killed local Map strain, we chose BALB/c mice as the experimental model, and use a prime-boost strategy with a MVA85A and challenge with a virulent local strain.
MATERIALS AND METHODS: 5 groups of 15 female BALB/c mice 7 weeks of age were used.
The groups were: A) unvaccinated control, B) 2 doses of the commercial vaccine Silirum® (CZ Veterinaria S.A.), C) 2 doses of the inactivated local strain, D) 2 doses of inactivated local strain with a boost of MVA85A and E) single dose of the inactivated local strain and the boost of MVA85A. The local strain evaluated as a vaccine candidate was prepared by removing bacterial clumps with 25G needle passages, heat inactivated and suspended in PBS at a concentration of 2.5 mg dry pellet /ml with incomplete Freund_adjuvant. 0.2 ml of the Silirum® vaccine or the local strain was injected subcutaneously, and MVA85A was inoculated intraperitoneally at a concentration of 5.6x105 PFU/ml. The different doses were administered 15 days apart. One month post vaccination, the animals were challenged with a virulent strain at a concentration of 1x108 CFU/mouse intraperitoneal. A pre-challenge blood sample was taken for the measurement of serum total IgG and isotypes (IgG1, IgG2a) by ELISA. 6 and 12 weeks post-challenge the animals were sacrificed and spleen was processed for CFU counting and cytokines quantitation.
RESULTS: The groups D and E which received the prime-boost scheme resulted in CFU counts significantly lower than the unvaccinated control group, and also lower than the vaccinated groups without MVA85A boost. Groups receiving 2 doses of vaccine (with or without MVA85A boost) had significantly higher levels of total IgG compared to a single dose group and the RATIO IgG2a / IgG1 was < 1 for all groups.CONCLUSION: The vaccination scheme with the local strain and the MVA85A boost resulted in better protection than the commercial vaccine in the murine model.

Keywords :
Vaccine, MVA

P-05.2

MYCOBACTERIUM AVIUM SUBSP. PARATUBERCULOSIS AND EBV PEPTIDES ARE RECOGNIZED IN SERA AND CEREBROSPINAL FLUID OF MS PATIENTS

Leonardo A Sechi[1], Giuseppe Mameli, PhD[1], Eleonora Cocco, MD[2], Jessica Frau, MD[2], Maria Giovanna Marrosu, MD[2]

[1]*Dipartimento di Scienze Biomediche, Sezione di Microbiologia e Virologia, Università di Sassari, Viale San Pietro 43b, 07100 Sassari, Italy.* [2]*Centro Sclerosi Multipla, Dipartimento di Sanità Pubblica Medicina Clinica e Molecolare, Università di Cagliari, Via Is Guadazzonis 2, 09126 Cagliari, Italy.*

Mycobacterium avium subsp. paratuberculosis (MAP) and Epstein-Barr virus (EBV) epitopes elicit a consistent humoral response in serum of multiple sclerosis patients, but the cross reactivity against the homologous myelin basic protein (MBP) and human interferon regulatory factor 5 (IRF5) has not been searched within the Cerebral Spinal Fluid (CSF). We evaluated in sera and CSF of patients with MS and with other neurological diseases (OND) the humoral response against EBV/MAP peptides and the IRF5/MBP.Our data showed that EBV and MAP peptides are able to induce a specific humoral immune response in MS patients compared to OND controls both in serum and in CSF. An intrathecal specific synthesis of IgG against MBP and their EBV and MAP homologous as indicated by the antibody index was observed in MS patients. The humoral response against EBV, MAP, MBP and IRF5 was significantly higher in MS patients compared to OND both in serum and in CSF. The higher presence of antibodies against MBP and their MAP and EBV homologous in CSF during relapses suggests a possible role of the pathogens in enhancing inflammation.

Keywords :
MAP, EBV, MS, CFS

P-05.3
Mycobacterium avium subsp. paratuberculosis in intestinal biopsies: risk factors

Espeschit, I.F.[1]; Carvalho, I. A.[1]; Oliveira, J. F.[1]; Souza, M.C.C[1]; Valêncio, A[1]; Ferrari, M.L.A.[3]; **Faria, A.C.S**[1]; Moreira, M.A.S.[1]

[1]Bacterial Diseases Laboratory, Section of Preventive Veterinary Medicine and Public Health. Universidade Federal de Viçosa-UFV. PH Rolfs Avenue, Campus Universitary, Zip Code 36570-900. Vicosa, Minas Gerais state, Brazil. [2]Department of Economics– Universidade Federal de Viçosa, UFV. [3]Alpha Institute of Gastroenterology- Universidade Federal de Minas Gerais- UFMG. Professor Alfredo Balena, Avenue, 110, room 208. Zip Code 30130-100. Belo Horizonte, Minas Gerais state, Brazil

Mycobacterium avium subsp. paratuberculosis (MAP) is the causative agent of Johne_disease or paratuberculosis, persistent gastroenteritis, which leads to progressive weight loss and that affects ruminants. In humans, MAP has been isolated from samples from patients with intestinal diseases such as Crohn_disease (CD), whose etiology and cure still unclear. The involvement of MAP in the pathogenesis of CD and other intestinal diseases is unknown, so it is important to detect the factors that favor or not the MAP infection in humans. A case-control unpaired was carried out by a questionaire from the population of patients with inflammatory bowel disease who underwent a colonoscopy exam, between the years 2009 and 2011 at a referral center in gastroenterology, located in the Hospital das Clinicas, Universidade Federal de Minas Gerais (HC-UFMG), Belo Horizonte-MG, Brazil. Eight of the 148 patients had positive intestinal biopsy for the presence of MAP DNA. It was found to be positively associated with the consumption of unpasteurized dairy products (OD = 13.39, CI = 1.57 to 298; p <0.01), history of intestinal diseases among family members (OD = 8.42; IC = 1.44 to 63.50; P <0.01). The study confirmed that that the consumption of informal milk and history of intestinal diseases among family members, are risk factors for having a positive intestinal biopsy sample for MAP.
ACKNOWLEDGMENTS We acknowledge the financial support given by the institutions: Fundação de Amparo à Pesquisa do Estado de Minas Gerais (Fapemig), Coordenação de Aperfeiçoamento de Pessoal de Nível Superior (CAPÈS) and Conselho Nacional de Desenvolvimento Científico e Tecnológico (CNPq). Maria Aparecida Scatamburlo Moreira is also individually supported by CNPq.

Keywords :
PCR, questionnaire paratuberculosis, milk, Crohn_disease

P-05.4
Effect of age at vaccination on Map shedding: a ten -year follow-up

Miriam Serrano[1], Elguezabal N.[1], Plazaola J.M.[4], Lauzurika J.[2], Ocabo B.[3], Geijo M.V.[1], Molina E.[1], A. Sevilla I.[1], Arrazuria R.[1], Alonso-Hearn M.[1], Juste R.A1.[5], Garrido J.M[1]

[1]NEIKER-Tecnalia, Basque Institute for Agriculture Research and Development, Derio, Bizkaia, Spain. [2]Diputación Foral de Araba. Araba. Spain. [3]Diputación Foral de Bizkaia, Bizkaia, Spain. [4]Diputación Foral de Gipuzkoa, Gipuzkoa, Spain. [5]SERIDA, Department of Agriculture of the Regional Government of the Principality of Asturias, Deva, Asturias, Spain.

Since 2005, 10 years have passed from the beginning of the field trial on paratuberculosis vaccination in the Basque Country. Previous data collected during this vast follow-up have been communicated in preceding colloquia. Here, we have included the data obtained over the last 3 years. The results obtained so far, are highly satisfactory for both: farmers and regional animal health authorities. Currently 21 vaccinated (VH) and 5 control (CH) herds submitted to a Test & Cull strategy, are included.Here, we report the differences obtained on Map fecal shedding and on tuberculosis Cervical Comparative Skin Test (CCST), related to age at vaccination time. At the moment of joining the trial, the whole (VH) herd was vaccinated with a commercial inactivated vaccine (SilirumTM, CZ Veterinaria, Spain), as well as all the calves introduced as replacement afterwards within their first 3 months of life (YV). Fecal samples from animals older than 24 months were taken and tested by rtPCR during the first herd sampling and once-yearly from then on. Simultaneously, the CCST was carried out according to European legislation.At this moment, 10924 CCST readings and 10521 faecal PCR results have been compiled. The initial fecal PCR prevalence for the CH and VH, reached 10.61% and 13.14% respectively. The shedding rate declined immediately after the first sampling. The prevalence for the annual samplings between 2012 and 2014 in animals vaccinated when older than 6 months (AV) was 6.5% (n=962) for the CH and 3.8% (n=684) for the VH, (p=0.019) and 0.39% (n=254) and 1.74% (n=518), (p=0.174) respectively for YV. Significant differences in fecal shedding according to the age at vaccination were observed (p<0.001). The overall frequency of positive reactions in the CCST (0.097% in YV vs. 0.3% in AV) is not affected by the age of vaccination (p=0.063), although the prevalence indicated a more advantageous trend when vaccinated under 6 months.In conclusion, vaccination at both ages showed different effectiveness related to fecal shedding, and although no significant reduction is observed in CCST, the results showed that the frequency of reactors in YV is lower than in AV group.

Keywords :
Map vaccine, shedding, cattle, tuberculosis

P-05.5
OPTIMIZED SURVEILLANCE FOR PARATUBERCULOSIS

Carsten Kirkeby[1], Kaare Græsbøll[1], Søren Saxmose Nielsen[2], Tariq Halasa[1], Nils Toft[1], **Gregers Jungersen**[1]

[1]*National Veterinary Institute, Technical University of DenmarkKaare.* [2]*Department of Large Animal Sciences, University of Copenhagen, Denmark*

Regional and national monitoring programs for paratuberculosis can be used to estimate the prevalence in endemic areas and surveillance in disease-free areas. It is important to monitor the prevalence before, during and after implementation of control measures such as vaccination or risk mitigation strategies. The design of such a monitoring scheme should be optimized to target the animals that are most sensitive to the test used, or at least reflect the variation in sensitivity.,
Due to the chronic nature of the infection, the test sensitivity of most ELISAs for Mycobacterium avium subsp. paratuberculosis (MAP) infection increases with the age of the tested animal. We can use this knowledge to find the optimal target groups for surveillance by estimating the mean sensitivity in a herd based on the age distribution of the animals, and by considering the effective lifespan of an animal. We name this measure the Mean Effective Sensitivity (MES). In this study we used a dataset of 4295 Danish dairy herds and 4078 beef cattle herds to estimate the mean sensitivity of a commercial MAP specific ELISA (ID Screen from ID-Vet) within each herd. Assuming that the sensitivity for milk and serum ELISA is the same, we then divided the herds into different groups based on age and herd type. We found that the general MES was about 34% for all dairy herds. However, when only testing animals > 2.0 years of age in the dairy herds, the MES increased to 60%. For the beef cattle herds, the MES was 30% when testing all animals, but increased to 65% when only testing cattle at least 2 years of age in herds with more than 25 animals that were at least 2 years of age. Moreover, we observed that seasonal calving has the largest effect on MES in beef cattle herds compared to dairy herds. These results are useful for defining the most optimal groups of herds for cost-effective targeted surveillance for MAP infections.

Keywords :
Paratuberculosis, decision making, monitoring, control

P-05.6
EFFECTS OF VACCINATION AGAINST PARATUBERCULOSIS IN A GOAT DAIRY FLOCK: A TWO YEARS FOLLOW-UP

Royo M[1], Fuertes M[1], Fernández M[1], Castaño P[1], Ferreras MC[1], Benavides JV, Pérez V[1]

[1]*Dpt Sanidad Animal, Instituto de Ganadería de Montaña (CSIC-ULE), Facultad de Veterinaria, Universidad de León, León, Spain*

In order to investigate the peripheral immune response associated with vaccination against paratuberculosis and their duration, as well as its effects upon animal losses, three groups of goats of different ages: A.- 1.5 months (35 animals), B.- 5 months (82 animals) and C.- adults (82 goats), were monitored in a commercial dairy herd of Murciano-Granadina breed formed by 199 adult animals managed in intensive system. Approximately half of the animals from each group were kept as unvaccinated controls, while the others were vaccinated with an inactivated vaccine (Gudair®). During two years after vaccination, and every three months, blood samples were taken from all the monitored goats.
Peripheral humoral and cellular immune responses were evaluated by indirect ELISA and IFN-γ release assay. All the casualties, regardless the reason, were registered and a number of them analysed by pathological methods.
Vaccination induced an intense cellular peripheral immune response, measured by IFN-γ production in stimulated blood that reached the maximum at 6 months post-vaccination (mpv). It progressively decreased and did not show significant differences with the control group at 15 mpv. The response was significantly higher in the group B than in the rest.
Similarly, antibody response reached the highest values between 3-6 mpv and was significantly intense in group B only in some samplings. Goats with lesions related to paratuberculosis were detected only in group C, and they appeared in lower numbers among the vaccinated animals. Moreover, a significant decrease in the total number of animals culled for any reason, related and unrelated to paratuberculosis, was observed in the three monitored groups among vaccinated animals, suggesting a general beneficial effect of paratuberculosis vaccination.

Keywords :
Vaccination, protection, goat, follow-up, immune response

P-05.7
A NOVEL APPROACH TO IDENTIFY THE COMMERCIAL DEER FARM AT HIGH LEVEL OF JOHNE'S DISEASE INFECTION IN NEW ZEALAND

JunHee Han[1], Peter R. Wilson[1], **Cord Heuer**[1]

[1]*Epicentre, Institute of Veterinaty, Animal and Biomedical Sciances, Massey University, New Zealand*

Johne's disease (or Paratuberculosis) in deer, caused by infection of Mycobacterium avium subspecies paratuberculosis (MAP), is granulomatous enteritis inducing diarrhea, weight loss and even death. In order to decrease the burden of the disease in deer industry, Johne's Management Limited (JML) was established by the industry stakeholders in New Zealand. The purpose of the program is to establish a syndromic surveillance program that monitoring the level of enlarged mesenteric lymph node (EML) in farmed deer at slaughtering process in Deer Slaughter Premises (DSP). The aims of this study were; 1) To describe longitudinal EML pattern in deer in New Zealand, and 2) To develop a novel way to identify the farms with high-level of Johne's disease infection using the data from JML program. This study described the longitudinal EML prevalence of commercial farmed deer in New Zealand by using linear regression and multiplicative SARIMA model. Overall national EML prevalence was fluctuating at 1 % with seasonal variation between 2007 and 2014. This study also developed a novel method to identify the possible outbreak source among the surveillance elements by comparing the longitudinal pattern of each element with the baseline. Predicted value of linear regression and SARIMA model with plausible variation were used for estimating the baseline, showing statistically similar results of identification. Though baseline estimation could be variable, overall scheme of detecting possible outbreak source in this study would be applicable in different types of data as well. Finally, this study identified the farms at high level of Johne's disease infection using the novel method. This study concluded that around 13 % of commercial deer farms in the country are at high level of infection. For the further study, however, more precise estimation on the level of Johne's disease infection is required by incorporating other information, such as temporal pattern of exceeding the baseline for each farm or geographical location of those farms.

Keywords :
Johne_disease, Deer, Syndromic surveillance

P-05.8
SHEEP BREEDS CONFIRMED WITH OJD IN NSW– REVIEW OF 779 LABORATORY SUBMISSIONS 1981-2006

Ian J. Links[1]

[1]*Graham Centre for Agricultural Innovation, Wagga Wagga NSW Australia 2650*

Introduction : While there is strong anecdotal evidence that Merinos are more susceptible to OJD than Crossbred/British Breed flocks, there are few published reports in the literature (Morris, Hickey et al. 2003). The results of field investigations for OJD recorded in the NSW veterinary laboratory system LABSYS database from 1981 to 2006 were reviewed to ascertain the Breed of sheep involved. Submissions "confirmed with OJD" were identified based on the coded laboratory findings recorded at the time the laboratory report was finalised.
Results : Laboratory submissions from sheep properties satisfying the criteria for "confirmed with OJD" were included in the present study if they were coded: JD positive (cases prior to 1993), JD histopathology positive or JD histopathology suggestive and JD culture positive. In cases where a specific Crossbred/British Breed was recorded, the breed, property and final diagnosis was checked on the original report. A total of 779 submissions were identified as "confirmed with OJD". After culling 125 "unknown" and 12 "mixed" Breed, the 642 remaining submissions were categorised as: Merino 582 (90.7%); Crossbred 32 (5.0%); 1st Cross Merino 10 (1.6%); Dorset 5 (0.8%); Border Leicester 3 (0.5%); Poll Dorset and Corriedale 2 (0.3%); Bond Corriedale, Suffolk, Texel cross, Wiltshire Horn, Perendale and Gromark 1 (0.2%). All 18 defined Crossbred/British Breed cases were from separate properties.
Conclusion : While the great majority of laboratory submissions "confirmed with OJD" were reported as Merino (90.7%) reflecting the dominant breed in NSW, there were 50 (7.8%) reported in the category Crossbred/British Breed, as well as a wide range of individual breeds recording at least 1 detection. These cases do not reflect all the confirmed cases of OJD in NSW over the period 1981-2006 as some submissions may not have been coded and the LABSYS database was progressively replaced by a LIMS database from the early 2000's.

Keywords :
OJD, ovine Johne_disease, sheep, laboratory confirmation, Australia, paratuberculosis, sheep breed, Mycobacterium avium subsp. paratuberculosis

P-05.9
Prevalence of ovine paratuberculosis in Lazio and Tuscany

Luigi De Grossi[1], Davide Santori[1], Alessio Gelli[1], Bruno De Santics[1], Antonino Barone[1], Erminia Sezzi[1], Marcello SalaIstituto[1]

[1]*Istituto Zooprofilattico Sperimentale del Lazio e Della Toscana, Italy*

Ovine paratuberculosis is a chronic disease of ruminants caused by Mycobacterium avium paratuberculosis. Epidemiological data are very rare although it is widespread.The aim of this work was to investigate about the prevalence of paratuberculosis in sheep in Lazio and Tuscany, central Italy regions.18,985 blood samples were analyzed by Elisa test and the reactive samples were successively processed to verification test.In the present work we analyzed 30 animals in each flock for a total of 15204 animals.Samples were collected randomly to get homogeneous groups of age and physical condition.Estimated prevalence using ELISA is influenced by the accuracy of the diagnostic test used and the comparison of data from different studies is needed to correct the apparent prevalence (AP) and so get true prevalence (TP).The prevalence of positive samples were 6.8% (8% Lazio, Tuscany 5.4%), with minimal observed prevalence of 1.7% and a maximum of 14.5% and the positives flocks were 53,8%.These results are according to the literature and are important for a future National surveillance Johne 's disease program for sheep.Nielsen et al., (2008); Hermon - Taylor J. & Bull T. (2002); Hermon – Taylor J. (2001); Khol, J.L., Stein, B., Dreier, S., Baumgartner, (2006); Mendes, S., Boinas, F., Albuquerque, T., Fernandes, L., Afonso, A., Amado, A., (2004); Nielsen S.S., Gronbaek C., Agger J.F.,Houe H. (2002); Schreiber S., Rosenstiel P., Albrecht M., Hampe J., Krawczak M. (2005)

Keywords :
Paratubercolosis, Sheep, Survellance

P-05.10
Evaluation of Effectiveness of Long Term Use of Killed Whole-cell Vaccine Against Mycobacterium avium subsp paratuberculosis in 3 Dairy Cattle Herds

Elisabeth Patton[1], Nicole Aulik[2], Matt Dodd[3], Scott Wells[4]

[1]*Wisconsin Department of Agriculture, Trade and Consumer Protection.* [2]*Wisconsin Veterinary Diagnostic Laboratory.* [3]*Anderson Veterinary Service, MN.* [4]*University of Minnesota, School of Veterinary Medicine, USA*

Johne's vaccination was initiated in 3 dairy herds in Northwestern Wisconsin as part of Johne's disease control programs in 2003. The herds participated in a controlled clinical trial in which every other heifer calf was vaccinated. During the study, individual fecal cultures were performed on cattle. Results from the study demonstrated that Johne's vaccination significantly reduced the risk of vaccinates testing positive for Mycobacterium avium subsp. paratuberculosis (MAP) and a trend toward fewer clinical cases. Organism detection-based MAP tests are needed in vaccinating herds, since vaccine responses may alter antibody responses. Often herds that utilize Johne's vaccine as a part of a Johne's disease control program, do not test for MAP due to the cost of organism-detection based methods. As such, prevalence data is limited in herds using long term Johne's vaccination. Although individual animal testing in the dairies was discontinued after the study was completed, all three herds have continued on Johne's disease control programs that utilize regular risk management assessments and Johne's vaccination. Data on current infection prevalence using pooled fecal samples on subsets from 2nd and higher lactations within each herd will be presented.

Keywords :
Mycobacterium avium subsp paratuberculosis, Johne_vaccination, infection prevalence, dairy, cattle

P-05.11

Assessment of within-herd seroprevalence of Mycobacterium avium subspecies paratuberculosis infection in dairy herds based on pooled or environmental samples

Raphaël Guatteo[1], Clément Moriceau[1], Alain Joly[2], Christine Fourichon[1]

[1]BIOEPAR, INRA, Oniris, La Chantrerie, 44307 Nantes, France. [2] GDS Bretagne, Vannes, France

Pooled individual samples, boot swabs or liquid manure sampled in different location in the barn are frequently suggested as an easy-to-use alternative to sampling individual cows in order to estimate the within herd prevalence of Mycobacterium avium subspecies paratuberculosis (Map). However, very few studies conducted under filed conditions are available. The aim of this study was to assess the informative value of diagnostic methods (PCR or ELISA), applied to pooled fecal (according to parity) or environmental samples in different locations, alone or in combinations, for discriminating cattle herds depending on their seroprevalence. To reach this goal, the data from 316 dairy cattle herds were used. In each herd, cattle older than 24 months were sampled (faeces and blood). Pooled fecal and blood samples were composed of 5 individual samples for each type of sample, within each parity class. Each animal had to be included at least in one pooled sample. Pooled fecal samples were tested using qPCR and pooled blood samples were testes using ELISA according to manufacturer instructions. qPCR was also applied to boot swabs collected, based on a standardized process, in different locations (bedding area, bulk tank milk, milking parlor). Then, the informative value (Sensitivity Se; Specificity Sp) of each [sample x test] alone or in combination was calculated. To discriminate dairy herds with a high within herd seroprevalence (> 10,5 %), the most informative method was the combination of concomitant positive PCR results on bedding area and faeces pooled samples (>50% of positive pooled samples) originating from primiparous (Se = 0,53 ; Sp = 0,82). To discriminate dairy herds with very low seroprevalence (< 5,5 %), the most informative method was the combination of negative PCR result on bedding area and negative ELISA results applied on all pooled blood samples in both primiparous and second lactation (Se=0,67, Sp=0,86). The present findings support the use of pooled or environmental samples in day herds for identifying a lower cost herds with low or high Map-prevalence which could at mid-term facilitate the organization of safer trades of cattle between herds. This study was funded by GDS Bretagne, GDS France & UMR Oniris-INRA BioEpAR

Keywords :
Dairy cattle, pooled samples, environmental samples, herd status

P-05.12

Regional Differences in Johne's Disease Prevalence based on Environmental Culture and Bulk Milk Testing

D. Kelton[1], H. Barkema[2], C. Bauman[1], C. Pickel[2]

[1]University of Guelph, Guelph, Ontario, Canada. [2] University of Calgary, Calgary, Alberta, Canada

The first Canadian National Dairy Study (NDS) was completed in 2015 with an overarching objective to benchmark the health, productivity and management of the national dairy herd. The study included over 1,340 dairy farms (11% of all dairy farms in Canada), of which 46% had participated in a voluntary regional Johne's disease control program. Regional programs in Canada are based on either fecal culture/PCR of environmental samples, or cow/bulk tank milk ELISA tests, limiting the ability to compare herd-level prevalence estimates among regions. As part of the NDS, 375 farms in all 10 provinces were visited and both environmental fecal and bulk tank milk samples were collected for testing.
A composite manure sample was collected from three different areas on each farm: the breeding age heifer pen, the milking cow area (alleyways), and from the manure storage area (either the liquid manure pit or the manure pile). Each of these three samples underwent DNA-based (PCR) testing for the presence of Mycobacterium avium ssp. paratuberculosis. Bulk tank milk samples were collected either at the time of the herd visit by study personnel, or after the visit by accessing payment samples collected by milk transporters. Milk was tested with a commercial ELISA kit. Based on results of environmental fecal testing, the prevalence of test-positive farms was highest in Western Canada and Ontario (20%), moderate in Eastern Canada (12%) and lowest in Quebec (5%). Results of bulk tank milk testing are pending and will be presented for comparison. Recognizing that these herd level tests lack sensitivity and are likely only detecting herds with a higher within-herd prevalence of infection, the regional differences are of interest and may be related to herd size and/or housing systems. Data available from the NDS will be used to identify factors associated with these differences.

Keywords :
Prevalence, milk testing

P-05.13

New criteria to differentiate paratuberculosis vaccination from bovine tuberculosis infection in the Cervical Comparative Skin Test

Serrano M[1], Elguezabal N.[1], Urkitza A.[2], Geijo M.V.[1], Molina E.[1], Arrazuria R.[1], A. Sevilla I.[1], Vordermeier M.[3], Whelan P.[3], Juste R.A.[1,4], Garrido J.M[1]

[1]*NEIKER - Tecnalia, Basque Institute for Agricultural Research and Development, Derio, Bizkaia, Spain.* [2]*Veterinary Surgeon, Gernika, Bizkaia, Spain.* [3]*Department of Bovine Tuberculosis. Ahvla, Surrey, United Kingdom.* [4]*SERIDA, Department of Agriculture of the Regional Government of the Principality of Asturias, Deva, Asturias, Spain*

Although the efficacy of paratuberculosis (PTB) vaccination has been repeatedly demonstrated, its use has been restricted due to its interference with current tuberculosis (TB) diagnostic methods used in official eradication schedules. The goal of this study was to assess new differentiation of PTB vaccination from bovine TB infection in the Cervical Comparative Skin Test (CCST).Fifteen Friesian calves were selected from a feedlot based on IFN-γ release assay (IGRA) negative results through 3 consecutive samplings at weeks 0, 4 and 12 (W0, W4, W12). Ten of them were vaccinated with 1ml of an inactivated PTB vaccine at W0. After introduction into level 3 biosafety containment facilities, 5 vaccinated and 5 non-vaccinated calves were infected with 105 CFU of M. bovis by the intratracheal route on W16, resulting in 3 different groups: Map vaccinated-M. bovis infected, only Map vaccinated and only M. bovis infected. The CCST was carried out at W13 with Avian (APPD) and Bovine PPD (BPPD).Validation of the new criteria for Sensitivity (Se) was carried out in a herd of 120 Friesian cows unknowingly infected with M. bovis when vaccinated against PTB with a commercial vaccine (Silirum) and annually submitted to CCST during 3 years. Validation for Specificity (Sp) was carried out in over 9000 CCST on PTB vaccinated cattle from 26 herds throughout 9 years.Results were read and interpreted according to the standard criteria based on the absolute differences of increase in mm between both PPDs and an alternative one based on the calculation of the relative increase in percentage of each one (>100% and BPPDfflAPPD).In the experimental setting, the standard interpretation yielded a Sp of 100%, but a Se of 40% for vaccinated animals. Using the alternative criteria, Se reached 80% and Sp 100%. In field conditions, standard interpretation resulted in a Se of 23.9% and a Sp of 99.8% while the relative interpretation yielded a Se of 81.7% and a Sp of 99.8%.These results indicate that the reduction of Se in the CCST caused by PTB vaccination can be advantageously overcome with the relative interpretation.

Keywords :
Mycobacterium avium subspecies paratuberculosis vaccine, Mycobacterium bovis, differentiation, diagnostic new criteria

P-05.14

Differing Paratuberculosis dynamics in beef and dairy herds

Richard Sibley[1], Peter Orpin[2]

[1]*BVSc Hon FRCVS, Westridge Veterinary Group, Witheridge, Tiverton, UK.* [2]*BVSc MRCVS, Park Vet Group, Leicester, UK*

The risks of Paratuberculosis entering and spreading within cattle herds have been assessed by farmers and their veterinarians as part of their Johnes Disease management plans. This study demonstrates the big differences in the biosecurity and bio-containment risks of Paratuberculosis entering and spreading within beef breeding herds and commercial dairy herds. The different risks can explain the high prevalence of Paratuberculosis infection in large dairy herds and the relatively low prevalence in beef breeding herds.

Method: A web based herd health management tool (myhealthyherd.com) has been extensively used to engage both beef and dairy farms in the management of infectious disease in their cattle herds. Over 3000 cattle farmers have opted to use the system for the management of Johnes Disease in their herds, which has included the assessment of risks of entry of MAP and the spread of the disease within the herd. The farm specific risk assessments are completed and an overall risk score assigned allowing comparison of risks between dairy and beef breeding herds.

Results: Risk data from 3020 cattle herds (2537 dairy and 483 beef breeding herds) has been analysed. There is shown to be a very wide range of risks, but dairy herds have a significantly lower level of risk of Paratuberculosis entering their herds compared to beef breeding herds, but there is a very much higher risk of the disease spreading within dairy herds compared to beef herds if the disease becomes established (75% of dairy herds have the highest level of risk of spread compared to 31% of beef breeding herds).

Conclusions: The different disease dynamics of Paratuberculosis in beef and dairy herds is demonstrated by this study, and can help explain the differences in prevalence within beef breeding herds and dairy herds. The results emphasise the need for good biosecurity to keep the disease from entering herds that have a high risk of spread, as well as the challenges that face dairy farms where infection is already endemic. However, many infected beef herds are at little risk of the disease prevalence increasing due to the low risks of spread.

Keywords :
Paratuberculosis, Beef Breeding, Dairy, Prevalence, Biosecurity, Biocontainment

P-05.15

Bio-load of Mycobacterium avium subspecies paratuberculosis infection in human patients suffering with thyroid disorder in Agra region of North India using Indigenous ELISA and Real Time-PCR

Saurabh Gupta[1], **Shoor Vir Singh**[1], Kundan Kumar Chaubey[1], Manju Singh[1], Bjorn John Stephan[1], Sahzad[1], B. Lahiri[2], Prabhat Agarwal[3], Ashok Kumar Tyagi[4], Jyoti Misri[5], Jagdip Singh Sohal[6]

[1]*Microbiology Laboratory, Animal Health Division, Central Institute for Research on Goats, Makhdoom, PO - Farah, Mathura, Uttar Pradesh, India.* [2]*Pathology Laboratory, Hari Parwat, Delhi Gate, Agra, India.* [3]*Department of Medicine, S.N. Medical College, Moti Katra, Agra, Uttar Pradesh, India.* [4]*Division of Veterinary Public Health, Indian Veterinary Research Institute, Izatnagar, Bareilly, India.* [5]*Division of Animal Science, Krishi Bhavan, New Delhi, India.* [6]*Amity Institute of Microbial Technology, Amity University Rajasthan, Kant Kalwar, Jaipur, India*

Mycobacterium avium subspecies paratuberculosis (MAP) infection, the cause of Johne's disease is endemic in domestic livestock and leads to chronic granulomatous inflammation of intestines. Since MAP is excreted through milk therefore it is continuously entering human food chain from infected lactating cows, buffaloes and goats. Bacilli has been consistently reported in humans beings suffering with different kinds of autoimmune diseases such as Crohn's disease, Type 1 diabetes, Hashimoto_thyroiditis, Rheumatoid arthritis, Multiple sclerosis etc. Present study investigated the presence of MAP in human population suffering with thyroid disorders (confirmed by pathology laboratories using 3rd generation chemi-luminescent assays) using 'indigenous ELISA kit' and 'Real time based IS900 PCR' assay. Bio-load of MAP infection in human beings with thyroid disorder was 35.1% in serum samples (n=76) collected from pathology laboratories in Agra region of Uttar Pradesh. Blood samples (n=28) screened by Real time IS900 PCR, 32.1% were positive for MAP infection. The blood samples positive in PCR were cultured in a specialized HEY medium with mycobactin J earlier used by Zhang and others to investigate the presence of MAP in human population with thyroid disorders. Cultures results are awaited and incubated for 8 weeks for the primary isolation of MAP which would be further bio-typed to estimate inter-species transmission. In the absence of control programs in human population, the study indicated large scale exposure of human population to MAP in the Agra region of Uttar Pradesh. Study indicated urgent need to further investigate role of MAP in precipitation and progression of Thyroiditis in human beings and also on association of disease with respect to the genetic susceptibility to MAP infection on exposure through consumption of milk and milk products.

Keywords :
Mycobacterium avium subspecies paratuberculosis, thyroid disorders, Indigenous ELISA kit, Real Time IS900 PCR

P-05.16

Age at Vaccination and Estimated Vaccination Levels for OJD in the NSW High Prevalence Area 2000-2009

Ian J. Links[1]

[1]*Graham Centre for Agricultural Innovation, Wagga Wagga NSW Australia 2650*

IntroductionVaccination of sheep for ovine Johne's disease (OJD) with Gudair® commenced in the NSW High Prevalence Area (HPA) in January 2000. The number of vaccine doses sold (2000-2009) when compared with the adult sheep population provides a raw estimate of annual vaccination activity. However the vaccine is one dose for life and sheep may be vaccinated as lambs (<12mths of age), hoggets (12-24mths of age) or adults (ffl 2 years). To provide a better estimate of "herd immunity" the "cumulative" percentage of adults currently vaccinated was determined. It was assumed that: a) All vaccinated sheep were culled at the end of their 4th year; b) Those vaccinated as lambs contributed to the vaccinated adult flock after 2 years and hoggets after 1 year, and c) All sheep vaccinated as adults were vaccinated at 2 years of age. ResultsThe number of flocks and sheep in the HPA declined by approx. 50% from 11,840 flocks and 12.1m sheep in 2000 to 6,140 flocks and 5.6m sheep in 2009. Total vaccine sales in the HPA from 2000-2009 were 10.9m doses, with 84,000 in 2000, peaking at 1.73m in 2003. Vaccine doses sold annually as a percentage of the adult population rose rapidly from 3.3% in 2001 to 8.7% in 2002, before plateauing at ~16% (range 12.7-17.1%) from 2003-2009. From 2000-2002, the age at vaccination was attributed to the following categories: 75% lambs (<12 months of age), 17% hoggets and 8% adults. In contrast, from 2007-2009 vaccine sales in the HPA were attributed to 94% lambs (<12 months), 2% hoggets and 4% adults (2003-2006 assumed similar). The "cumulative" (rolling 3 year) adult vaccination status in the HPA reached an estimated 6% by the end of 2003, ~16% by 2004, ~45% by 2006 and ~70% in 2009. Abattoir monitoring over the same period confirmed a progressive fall in the percentage of animals with lesions attributable to OJD from 2.4% in 2000 to <1% from 2005. ConclusionWhile the primary message was to vaccinate as lambs, many producers, particularly those suffering heavy losses, were found to have vaccinated their whole flock or older age groups. It is likely that this contributed to a more rapid increase in "Herd Immunity" both within individual flocks and the general sheep population of the HPA. This was associated with a more rapid decline in population prevalence of OJD than might otherwise have been observed.

Keywords :
OJD, vaccine, Gudair, sheep, Mycobacterium avium subsp paratuberculosis, ovine Johne's disease, abattoir monitoring

P-05.17
Experiences with developing a commercial solution to Johne's control in the UK

Peter Orpin[1,2], Richard J. Sibley[3]

[1]*Park Vet Group, Whetstone, Leicestershire.* [2]*Special Lecturer, Nottingham University, Nottingham, United Kingdom.* [3]*Westridge Veterinary Group, Witheridge, Tiverton, UK*

In 2010 Dairy UK (milk processor representative body) formed a Johnes Action Group to tackle JD in the UK dairy sector. A collaborative JD engagement programme was developed to engage the dairy industry and veterinarians with the effective control of Johne's disease. The early outputs of the group were focused on establishing coherent messages for inclusion in vet and farmer training, standardised surveillance (targeted 30 cow milk ELISA screen) combined with a commitment to robust risk assessment and management of JD. As a parallel development a commercial web based health planning tool (www.myhealthyherd.com) was designed linking the farmers, vets, monitoring organisations and labs with 2537 farmers completing risk assessments. A specific Johne's manager module within the program facilitated structured risk assessments of disease entry and spread to be completed. A prevalence prediction graph was created using a combination of a targeted milk ELISA sampling of 30 high risk cows and risk data to prompt action from the farmers.In the last 5 years over 300 farmer meetings to 2500+ farmers have been delivered via their milk processors with a further unknown number delivered by private vets. Over 350 vets have been educated in the principles of JD control. In 2015 the engagement plan was enhanced by the development of the National Johnes Management Plan (NJMP) to engage the processors and their farmers to commit to effective JD control. The first phase of this programme focuses on testing for JD, risk assessments and choosing one of 6 control strategies . By September 2016 NJMP has progressed through further financial support from milk processors whereupon 77% of the GB milk supply will have complied with the programme. A webinar 4 stage module of vet training is planned for 2015/16 to create a list of approved JD vets. The success of the program has been driven by creating commercial drivers for engagement for all participants in the NJMP.

Keywords :
Risk, Johne's control, myhealthyherd

P-05.18
Vaccination against Paratuberculosis in sheep: methodological developments and evaluation of efficacy

Mathevon Y.[1], Lafort M-P.[2], Tasca C.[1], Galan E.[3], Falguieres R.[3], Foucras G.[1,2], Corbiere F.[1,2]

[1]*INRA, ENVT, UMR 1225, IHAP, 31076 Toulouse, France.* [2]*University of Toulouse, INP, ENVT, 23 chemin des Capelles, 31076 Toulouse, France.* [3]*CAPEL Ovilot, 237 avenue Pierre Semard, 46000 Cahors, France*

The absence of a perfect diagnostic test towards paratuberculosis and the large flock size make the test and cull control strategy most often non-applicable or unsuccessful in infected sheep holdings. In France, the use of a killed vaccine (Gudair®) was allowed for sheep in 2009 and vaccination was implemented in highly infected flocks. The aim of our study was to assess the vaccine effects on serological response and faecal excretion in some vaccinated commercial flocks in comparison with non-vaccinated control flocks.Nineteen flocks with endemic paratuberculosis were involved in the study. By 2009 onward, 3 to 6 month-old replacement ewe-lambs were vaccinated every year in nine of these flocks. Based on pre-vaccination serological surveys, the apparent prevalence towards paratuberculosis was significantly higher in the nine flocks that implemented vaccination programs than in the ten flocks that did not (median= 12.2% vs 2.9%, Wilcoxon Signed rank p=0.01). During 2014, individual serum and faecal samples were collected on 150 two to five year-old ewes in each vaccinated flock (total = 1617), and on 100 ewes aged from 2 to 3.5 years in non-vaccinated ones (total = 934) and analysed using commercial ELISA (ID Screen Paratuberculosis Indirect, ID.Vet) and qPCR (Adiavet, Biomerieux), respectively. In vaccinated ewes both the serological prevalence (over 95% in all flocks) and S/P ratios were high, without any decrease with age. Serology was therefore deemed useless to follow up new infections in vaccinated farms. qPCR outputs were analysed in R 3.3.2 using the qpcR package by fitting the fluorescence curve from each sample to a specific model. Based on dilutions of samples containing known concentration of the Map genome, we achieved a precise quantification of faecal excretion. Results showed that in all vaccinated flocks the apparent prevalence of faecal excretion was low (0% to 7%, overall 2.9%) compared to unvaccinated flocks (0% to 30%, overall 8.0%). The distributions of excretory levels were similar in both flock types, with rare heavy shedders. Our results reinforce the idea that even if beneficial in few years, long-term vaccination is necessary to reduce the risk of new infections.

Keywords :
Ovine, vaccination, qPCR, control plan

P-05.19

REGIONAL BIOSECURITY PLAN AREAS FOR OVINE JOHNE'S DISEASE IN AUSTRALIA

Lawrence C Gavey[1]

[1]*Principal Veterinary Officer (Animal Disease Containment), Biosecurity Queensland, Department of Agriculture and Fisheries, PO Box 102, Toowoomba, Queensland, Australia*

Abattoir monitoring of sheep in Australia shows contrasting regions in which prevalence and distribution of paratuberculosis are continuing to increase versus significant areas which maintain demonstrably low prevalence. These low prevalence areas have adopted a coordinated set of standards that protect against spread of infection into and within the areas. As Australian management of paratuberculosis is shifting from regulatory protection of low-risk sub-populations to industry management that minimises the impacts of infection, the coordination of protection against risk is being progressively implemented by industry groups, with reducing levels of government involvement. Industry is assuming shared responsibility for biosecurity outcomes, and compliance with protective standards is at least as high under collaborative and voluntary responsibility as it has been under government regulation. Factors in the success of this non-regulatory model, particularly industry engagement and compliance and abattoir surveillance, are discussed. This encouraging model based on farm and regional biosecurity plans has application to a range of other sheep diseases such as footrot, brucellosis, lice and anthelminthic resistance, and to other livestock industries.
REFERENCES Links I, Evers M, Wright T, Barwell R. (2012) Evaluation of 108 abattoir traces for ovine Johne's disease to the New South Wales low prevalence area – a standardised approach. Proceedings 11ICP, Control Programs p.176Anon. (2013) New OJD Management Plan 2013-2018 http://www.ojd.com.au/australias-ojd-program/new-ojd-management-plan-2013-2018/Anon. (20143) Regional Biosecurity Plans 2013-2018, available from Australian sheep industries' paratuberculosis website: http://www.ojd.com.au/trading-sheep/regional-biosecurity-plans/
ACKNOWLEGDEMENTS The resolution of regional industry leaders and technical advisors in coordinating and maintaining regional biosecurity plans for Johne's disease in sheep, and the administrative support of Animal Health Australia, are gratefully acknowledged.

Keywords :
Regional biosecurity plan, ovine Johne's disease, sheep, Australia, industry management, non-regulatory

NOTES

NOTES

NOTES

NOTES

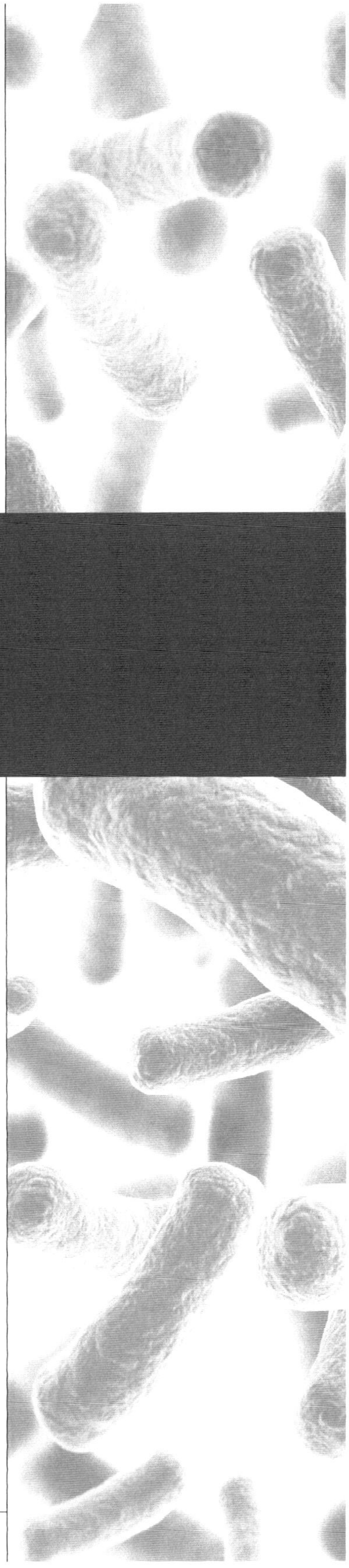

Author index

MAP

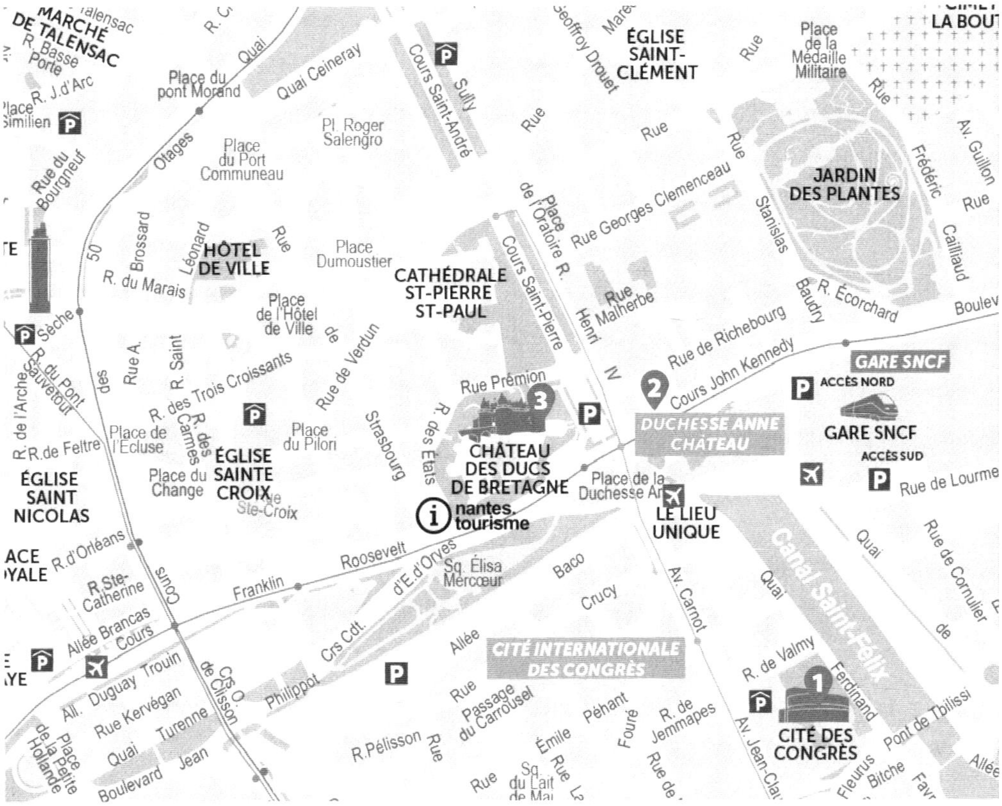

Tramway line n°1

Tramway line n°2

Tramway line n°3

Airport shuttle station

1 Conference Venue
La Cité Nantes Event Center
5, rue de Valmy
44041 Nantes, France

2 The Brady's pub
22, allée Commandant Charcot, Nantes
Sunday, June 19th
GET TOGETHER DRINK

3 Château des Ducs de Bretagne
4, place Marc Elder, Nantes
Thursday, June 23rd
COLLOQUIUM DINNER

UNDER THE PATRONAGE OF

PLATINUM SPONSORS

SILVER SPONSORS

BRONZE SPONSORS